David W. Pearson & Ron Bommarito

I0799361

ANTIQUE MINING EQUIPMENT AND COLLECTIBLES

2nd edition updated and expanded

Schiffer Publishing Ltd
4880 Lower Valley Road • Atglen, PA 19310

Library of Congress Control Number: 2015955813

Type set in Windsor BT/Dutch801 Rm BT
ISBN: 978-0-7643-5089-4
Printed in China

Published by Schiffer Publishing, Ltd.
4880 Lower Valley Road
Atglen, PA 19310
Phone: (610) 593-1777; Fax: (610) 593-2002
E-mail: Info@schifferbooks.com
Web: www.schifferbooks.com

CONTENTS

PREFACE

The candlestick holder on the rock demonstrates its use in a hard-rock mine. The green or blue color is high-grade copper.

CAUTION!

Please read this section before using this book. Words of caution are always imperative to the collector and dealer of antiques, however, this is imperative in the field of old mining collectibles. This field is highly specialized and the neophyte may suffer real financial losses in the process of buying or selling items. There have been collectors who have specialized in mining collectibles for years and only recently recognized some items for what they really are. The field is a large one where the knowledgeable buyer can buy and sell well.

With those thoughts in mind, we have written this book with the hope that it will be of help to beginning and serious collectors and dealers alike. In the process of accumulating the information, we have had the privilege of looking at and handling several of the finest collections of small mining equipment in existence. Our knowledge of the rarity and value of many of these items is based on lifelong accumulations of items in the field.

In some respects, being fortunate to live in or close to historic mining areas is a real advantage, since it allows us to observe a large cross section of the collecting field. In one respect it is a disadvantage, since regional differences in the market can cause prices to be slanted one way or the other. In some geographical areas mining artifacts are looked upon as oddities and often go wanting for buyers.

Condition

Further cautionary words are necessary concerning condition, and we will use some standard terms in the collecting field to help put our readers on a common playing field. The highest, most desirable condition is MINT IN THE BOX (MIB). This is the ideal condition and not often found. It is also the condition to be the most skeptical of and the most wary. A new condition carbide lamp occasionally does turn up and it is pretty obvious when one is found. It is the exception to find an absolutely new miner's candlestick, as these were almost always used and used hard. Beyond the usage, it is all too easy to fake a hand-forged candlestick and we have observed fakes in this field.

EXCELLENT CONDITION would be described as fine condition with no damage and may have been used or fired. Finish would be 98%. VERY GOOD condition would be described as used with minor blemishes but still completely functional. Finish would be 80 to 95%. GOOD condition is considered to be normal wear and tear with blemishes and 60% or better finish and functional. FAIR condition would refer to items with minor parts missing and some blemishes with some finish remaining. POOR condition would have missing parts, replacement parts, or alterations, with no original finish remaining. ARTIFACT condition refers to a dug up item in generally poor condition with no function remaining.

For the purposes of this book, we consider our price ranges to include GOOD to EXCELLENT condition. On a scale of one to ten, our prices are based on an eight.

Location/Geography

Certain types of items will be found more readily in some geographic regions than others. Here is where a regional difference will come into play in the pricing. Buying or selling from established dealers and collectors is usually good practice, but not a guarantee of quality in the mining collectibles field.

Books

We advise collectors to build a substantial library of books that illustrate and describe mining equipment. This is an area of collecting which we deal with in this book, but to be honest, there are many more books on the subject than we have room to include here. There are new books and old books and many in between, and they all have some utility to the collector. In fact, some newer books on specific items have themselves reached collectible status, such as the book on candlestick holders.

The scruffy looking white ore cart, $950. Round bottom "Mule dump" cart, $1,300. EC

ACKNOWLEDGMENTS

This book has been a long time in preparation and we want to apologize for the quality of a few of the photos. Simply put, we had to capture on film a couple of collections that were too large to move from out-of-the-way places in the California mother lode district. Some of the photos were taken under rather adverse conditions, but the items recorded were so rare and pristine in condition that the necessity of recording them for posterity outweighed the problems involved.

We would like to thank and acknowledge the input and collection of Errol Christman, who is or was the owner of one of, if not the largest collection of miners' carbide lamps in existence. Sadly, many of the items in his collection—photographed in 1996 and 1997—are now elsewhere, and to duplicate the original feat of recording these items would be impossible.

Similarly, we recorded the collection of the Smith brothers, which was large and inclusive of many of the cap tins and explosive items in this book. Sadly, both men are now gone off to the last roundup in the sky and their collection has been broken up. Others who have contributed to the information in this book and to the photos and collectibles are: Ron Bommarito, Anthony Moon, Bud Uzes, Bob Muraldo, Herb Dick, and Roger and Connie Peterson.

Photos were all taken by author Dave Pearson (except as noted), using either an aging screw mount Pentax camera or a Minolta 350 and both Kodak and Fuji film. Film was processed by your friendly local Walmart and later by Carson Photo Supply, who did superb quality reproductions. All photos were taken using available light where possible or artificial light where necessary.

We also want to thank the staff of the Nevada State Museum in Carson City, Nevada, for allowing us to record and photograph a number of items that are not only very rare but unique. Without the help of Doug Southerland this would not have been possible. Finally, we want to thank Fred Holabird of Fred Holabird Americana, an auction house that specializes in antique mining goods and specializes within that area in stock certificates and paper items. His catalog is worth purchasing for the history continued therein and his address is listed on page 174.

Specific collections are noted as follows:

AM for Anthony Moon
BM for Bob Muraldo
BS for Bob Samay
BU for Bud Uzes
DK for Dennis Kurlander
DS for Doug Southerland
EC for Errol Christman
HD for Herb Dick
NM for Nevada State Museum
RB for Ron Bommarito
RP for Roger Peterson
RS for Robert Schroth
SB for Smith brothers

Unless otherwise noted, the photos were taken by Dave Pearson.

The "Klondyke Nugget Game" is very rare and dates from the turn of the nineteenth century. $1,000. EC

Chapter One

ADVERTISING

One does not usually speak of advertising in the same breath with mining, but the reality is a number of highly collectible items have been produced through the years that were only intended for advertising mining suppliers' products. Most often these items were intended to induce the miners or the companies they worked for to use a certain product. Sometimes these advertising items are small, as in the case of miniatures of tools or lamps, but more frequently the advertising took the form of liquor bottles or tobacco in one form or another.

Miniatures

In the case of miniatures these are now highly sought after, as they would have commonly been given to some clerk or boss's child and have long since been lost. The most commonly found items in this area are miners' safety lamps and tools. Some of the miniatures are true working replicas of the real thing and these will bring premium prices. More common are items that are mock-ups of the full-size tool.

Tobacco Tins and Other Tobacco-Oriented Items

It is relatively common to find tins of tobacco such as "Miners cut" in antique sales. These were a popular form of tobacco usage and provided the underground miner with tobacco he could roll into a smoke, put in a pipe, or just chew. If the mine where he worked did not allow smoking on the job—and most did not for fear of starting a fire—then a healthy chew got the job done. Cans will almost never be found with full contents or sealed, but outside condition is very important.

In our preparation for this book we also found a pipe made in the form of a black powder barrel. The pipe is made from some sort of reddish-brown material such as Bakelite and marked with the DuPont name and logo. It was obviously intended to be used with a hollow wooden stem, but we have found no other reference to this item anywhere.

Bottles

These are included in advertising, mainly since they were a marketing tool for liquor sellers to ply to the miners' trade. The contents were probably no better or worse than "Old Overshoe," but the thirsty miner did not care. The best bottles will have embossing or coloring like the infamous "Jim Beam" bottles of this century. More commonly the labels were printed and glued on. Condition is the real factor, as there are many, many bottles out there.

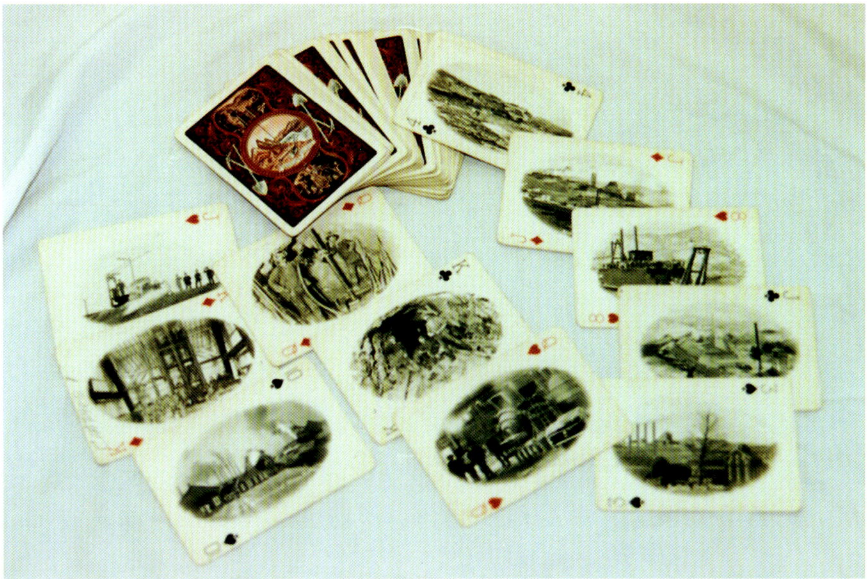

Deck of playing cards with mining scenes on reverse from Butte, Montana. $100 to $300 depending on condition. RB

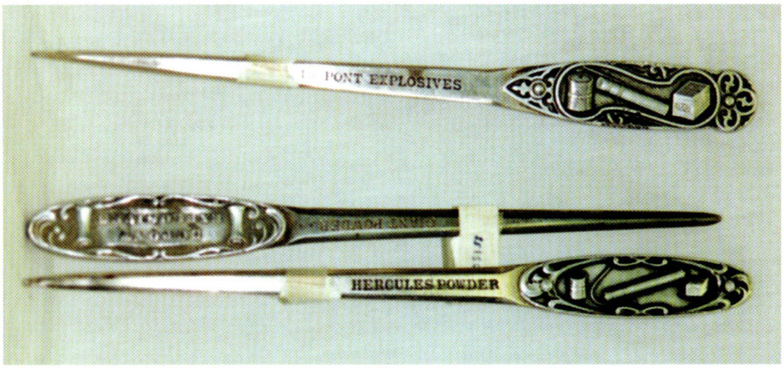

Three nickle-plated letter openers from powder companies. Given out as advertising. $65 to $135. RB

This grouping is a set of miniature miners' tools. They were probably intended to be used as salesmen's samples. $200 to $450. RB

The two items in the foreground are miniatures shown with a carbide lamp for scale. The helmet is an early pressed leather or fiber type, while the tin is a small version of a DuPont 25-pound black powder tin. $150 each. SB

Watch fobs were common for men who did not use wrist watches. The fobs were sold or given away by suppliers. This one is marked with the logo of Ingersol Rand (a heavy equipment supplier) and is made of copper. $50 to $75. RB

Brass token showing miners' tools given out in Butte, Montana—probably an advertising gimmick. $100 to $150. RB

Watch fobs are a bit difficult to value and can depend on the metal from which they are made, as well as overall condition. $35 to $100, each. EC

Miner's watch fob from Silver City, Nevada, which is part of the Comstock Lode area. This one is marked as belonging to Edward Hickey and is made of brass or copper. $75 to $125. DS

Bottle in the shape of a prospector. These are known to come in different sizes and believed to be made in Germany. $125 to $150. RB

A finely crafted miner's watch fob of brass with gilding depicting two miners using an air drill. $100 range. RB

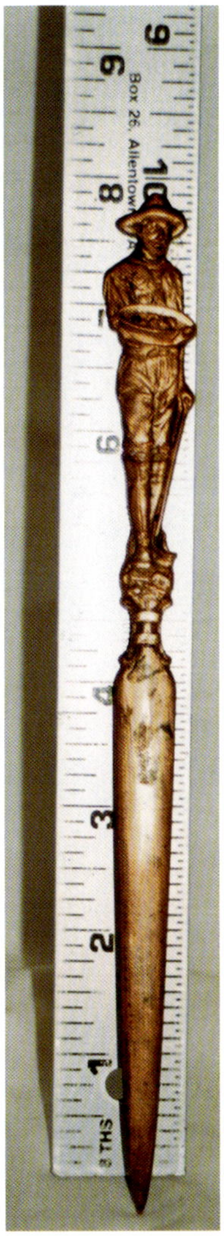

Solid copper letter opener in three-dimensional style. Date unknown. $75 to $100. RB

We're not sure what "stripped tobacco" was exactly, but it was supplied in a nicely made tin package with printed graphics. $200. RB

A collection of match holders. These were fairly common at one time, were waterproof, and usually made of stainless steel or nickel plated brass. $25 to $40. EC

Signs are not always the same. This is actually a stencil cut from sheet brass and came out of a mine in the Comstock Lode. It is dated very early in that district's time span. $900 to $1,000. RB

Not a sign, but actually a metal stencil for marking whatever needed marking around a mine or mill. This one comes complete with a couple of bullet indentations. Location of the mill was in the Hamilton, Nevada, mining district. $200 to $300. RB

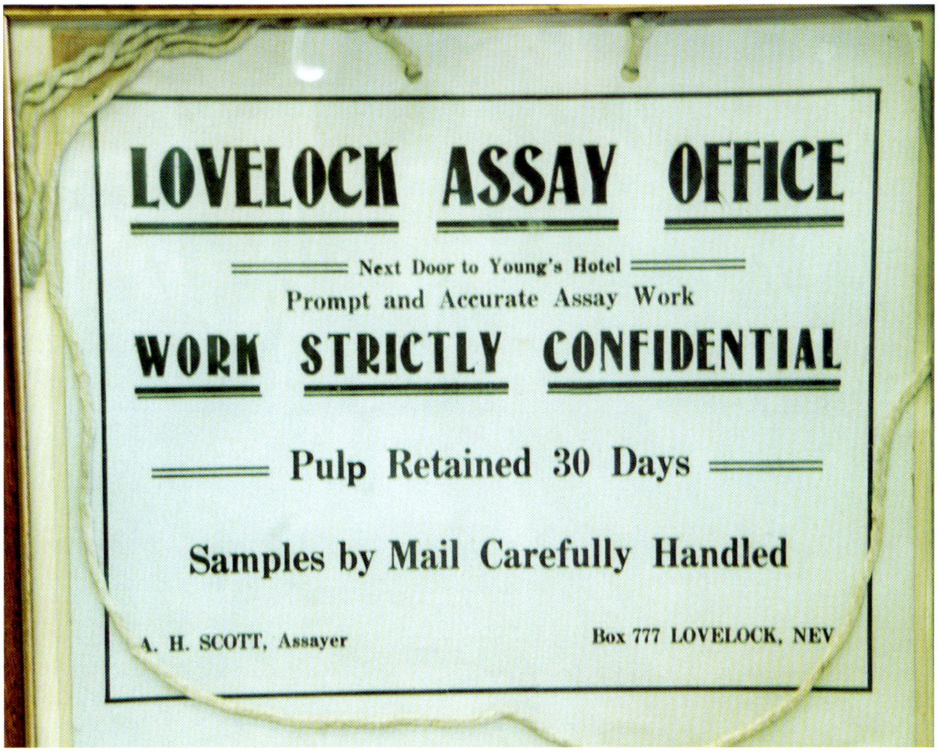

Another sign from A.H. Scott assayer, this one from a Lovelock, Nevada, address. $350 to $500. RB

A solid copper letter opener given out as advertising by the Phelps Dodge Corp, age unknown. The rock on which it rests is silver ore with a heavy inclusion of garnets. $100. RB

This sign from Goldfield, Nevada—rare by itself—is more so because it is painted or etched on tin and framed in an antique gold leaf frame. Circa 1905. $500 to $700. RB

Dating from about 1910, this advertising sheet poster is from Tonopah, Nevada. $400 to $550. RB

A small tobacco pipe (without the stem) made by DuPont for sale to miners. $125 to $175. RB

Copper money clip given out by Ingersoll Rand. $100. RB

A cute ash tray given out by Bear Brand powder products. Made in the shape of a miniature powder can, this is about three inches high and in like-new condition. $75. RB

Right:
A small grouping of items illustrated to display unusual collectibles. The miniature ore bucket has a blasting cap tube (empty) made by the Western Cartridge Co. The match safe, $50 to $60. The pencil holder, $100 range. The cap tube, $20. RB

A well-used enamel on steel sign advertising the products of the Giant Powder Company. $750 to $1500. RB

Matchbox cover of copper from the Trojan Explosives Company. $100 to $250 depending on the company. RB

An advertising poster for explosive items. $100 to $150. SB

Advertising display boards are highly collectible, even from companies who supplied other industries as well as mines. Wire cable was in heavy use by all mines. This is a Wickwire display board. $1,000. RP

A fine-condition label from a tobacco tin depicting an early miner. $75+. RB

This luggage tag was given out by the Globe Iron Works of Stockton, California. Globe was a major manufacturer of iron products for the western mining trade. $250. RP

Local beer was a staple of mining camps, as the product did not travel well. This label is from a long defunct brewery from Jackson, California, a major center of mother lode mining. No date. Valued at $20 by its owner. RP

Left:
Butte, Montana, was a major producer of copper for many years. This letter opener is marked from Butte, but there is no indication of its manufacturer. Heavy weight. $85. RP

DuPont was a major supplier of products for the mining industry and this 12" by 12" felt counter pad is in nice condition. $75. RP

Probably eastern US in origin, this cast iron ash tray was given out by the Kaier's Beer Co. $50 to $70. HD

The owner is unsure of the use of this small "pill box," but it is nicely crafted of a hard white material and labeled from the Dixon Crucible Co. $125. HD

This smallish cast and plated statue of King Coal was given out by the Blackhawk Coal Co. It was probably an advertising giveaway. $150 to $200. HD

Smoking tobacco from the Milwaukee, Wisconsin, firm of B. Leidersdorf Co. Not especially rare. The tin may be worth less than the paper container. $100 to $150. AM

CHAPTER TWO

ASSAY EQUIPMENT

In every mining camp there would have been a person who could perform a rudimentary assay on someone's ore that had just been dug from the ground. The difference between a good mining area and a dry wash is determined by the richness and type of the ore, and the first assayer in a camp who could give an honest and accurate determination of ore value was the one who usually got rich in the process.

The Western United States is today largely set out geographically because of the discovery of nearby high-grade ore bodies. Indeed, California would not be a part of the US if it were not for the mother lode district in the Sierra Nevada Mountains. The original gold rush was started with the discovery of large placer gold deposits. Placer gold is normally very high-grade gold found chemicaly free settled out in collection points in stream beds. The astute miners quickly came to the conclusion that a "mother lode" existed that supplied the streams with the gold dust they sluiced and panned out of the rivers.

It was not too many years after the initial discovery of 1848 that other miners were prospecting for more gold in an area not known for its gold: western Nevada. Actually then part of the Utah Territory, Nevada was put on the map with the discovery of very high-grade ore that looked like blue clay in an area that became known as Virginia (later changed to Virginia City), Nevada. The story tie-in to assaying is that initially the blue clay was discarded as trash by miners who were prospecting for free gold. One curious miner took a sampling of the clay to a better assayer in Placerville, California, and the rest, as they say, is history. A proper assay revealed that the clay substance was in reality almost pure silver. Subsequent testing also found significant quantities of gold in the ore.

Assayers' offices were initially set up in tents and later in small buildings. As they required a reliable source of heat, they were usually provided with a kiln or fire ring fed by charcoal and a bellows to reduce an ore sample to a "button" of metal that would then be tested by other chemical means to determine its make-up. An excellent book titled *The Mining Camps Speak* (Beth and Bill Sagstetter) not only illustrates how an assay office worked, but also has photos of an early Colorado operation.

All assay offices had to have some basic tools to accomplish their work, including a crusher of some sort, a heat source, an assortment of chemical bottles, a good set of scales, and books to help out with record keeping and reference material.

Assay Kits

The typical assay kit contained small containers of chemicals so the owner could do a field estimate on raw ore. The small containers were normally small jars with screw tops or glass stoppers. The value of these kits is directly related to the completeness and condition of the kit. They are commonly found missing certain items, such as the blowpipe used to provide oxygen to a small flame, usually an alcohol lamp and various bottles. The kits vary in size, but are typically compartmentalized and lined with fabric. The more use the kits had the more wear and tear they will show. Some kits are extremely high quality and quite complex.

Cupels and Cupel Makers

These are items that can be found, but most often are discarded as being trash. We have been fortunate to find quantities of used Cupels in one of our mining camp explorations. A cupel is simply a solid, cylindrical piece of material—usually made of bone ash—formed by the use of a mold and a press called a cupel maker. Every assay office had one and they are not common today, probably because most people do not know what they are. Cupels came in different sizes once made and were the non-flammable base where the assayer would heat a reduced sample of ore, then take the resulting button and weigh it for mass. They are similar to a crucible in usage but much smaller, ranging in size from ¾ inch to 1½ inch in diameter by 1½ inch high.

Chemical Bottles

Chemical bottles have changed little through the years, and the nicely molded bottles containing acid in my high school chemistry lab would have fit perfectly in the assay lab of any frontier mining camp. The earliest chemical bottles will have applied tops or "glob" tops and a fitted glass stopper. These glass bottles will normally only have value if they can be positively ascertained to have been used in a documented assay office. A cautionary note about old acid bottles: just because they are old does not mean they can't hurt you. Wash them in plenty of hot, soapy running water and use rubber gloves handling them until you are sure they are clean of all old residue. This is one item where cleanliness is proper and will not degrade the value, except for those with pasted-on labels. These should be cleaned very carefully to retain any handwriting on a paper label.

Crushers for Assaying

Some of the better collectibles found today are the various kinds of rock crushers used by assayers, ranging from the simple mortar and pestle—usually cast in iron—to a large hand-held tool called a muller. The muller was wielded like a large sledgehammer, but since it had a heavy curved head it was merely rolled over a rock sample until the pieces were either dust or small fragments. The muller would have been used on a heavy table, usually with a polished rock surface or a thick steel plate. The crushed sample was then put into the mortar and finely reduced before being chemically treated or cooked down with heat.

Later crushers of one type or other using rotary power were introduced that did the same task with less muscle power. Small electric crushers are still used in the assay process and by small independent miners.

An array of assay button molds all of cast iron. No makers noted. $50 to $250 each, depending on size and condition. RB

A rather crude but serviceable small rock crusher for use in assaying. Up to $400. RB

A rock "disc grinder" that would have been powered by another power source. The small rocks were introduced into the square opening and a rotating disk ground the ore to a fine grade. This would have been used in an assay office and was made by the Calkins Co. $450 to $600. RB

This odd-looking gizmo is a hand-charged air or gas pump used to provide pressurized air or flammable gas to an assay kiln. $200 to $300 range. RB

This type of mold is referred to as an assay mold and allowed the assayer to cast either gold or silver (or both) in the same mold. The ingot sizes are all different. The mold is made of rough cast iron with finished mold spaces. $100 to $200. RB

An impressive array of iron or steel mortars and pestles used for reducing ore to fine size for testing and assaying. $95 to $600. RB

A complete cupel maker with mold and a small mercury flask. The cupel maker, $150 to $200. The small flask, $300. RB

This strange-looking device was used to sort samples for assaying in an objective method. $150 to $200. RB

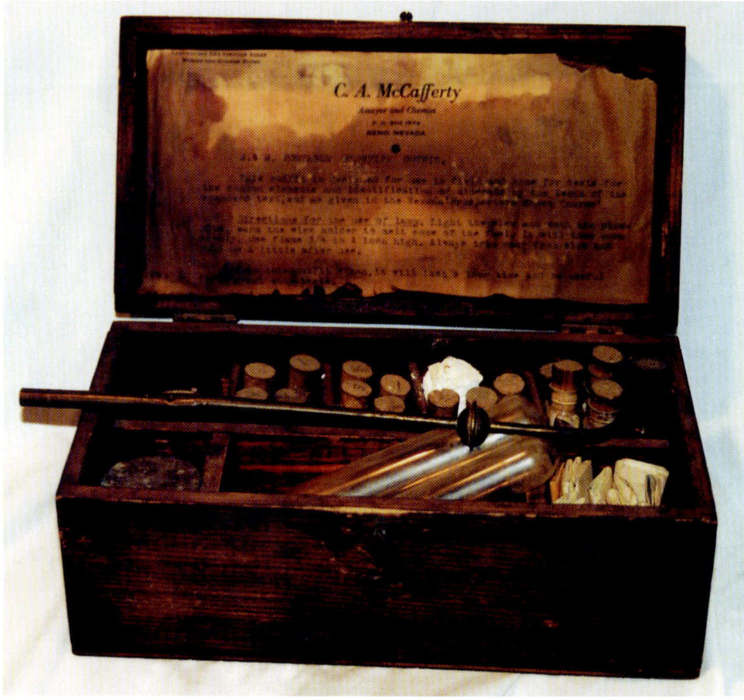

Assay kits of this quality and size are rather rare these days. This one is complete, including paper work. The long tube resting on the top is a blow pipe used to increase the heat of the flame source, usually a spirit lamp or alcohol lamp. $500 to $1,000, depending on completeness and condition. RB

Magnifiers in common usage are generally well made. $30 to $50 each. RB

Assayer's sample bag from Denver, Colorado. $45 to $100. RB

This large portable scale is finely made and in complete working condition. When assembled it stands approximately eighteen inches high. Complete with weights, $1,000. RB

Marked with a stamp reading "E.A. Martin & Co Assayers, Virginia Nev." (Virginia City), this assayer's ingot would contain both gold and silver and is about two ounces in weight. Rare and unusual, these would have been kept only to establish the quality of ore from a particular batch or area being mined. $2,000 and up. NM

These odd-looking "muffin trays" are actually button molds for use in an assaying furnace large enough to handle multiple samples. $75 to $200 each. RB

A small ingot from the Nevada State Museum collection that reads "First bullion produced at Tonopah, Nev. Tonopah mining Company. Nov 1903 (?) Fred J. Seiders." High-quality silver—now oxidized on the surface—accounts for the black color. This type of collectible is nearly impossible to value, as it is truly one of a kind. $2,500 and up. NM

Slightly different, this is a tray for holding cupels while in an assay furnace. Marked by Calkins, Los Angeles. $200. RB

A mechanical separator pan (also called a "splitter") used in assaying. $100 to $125. RB

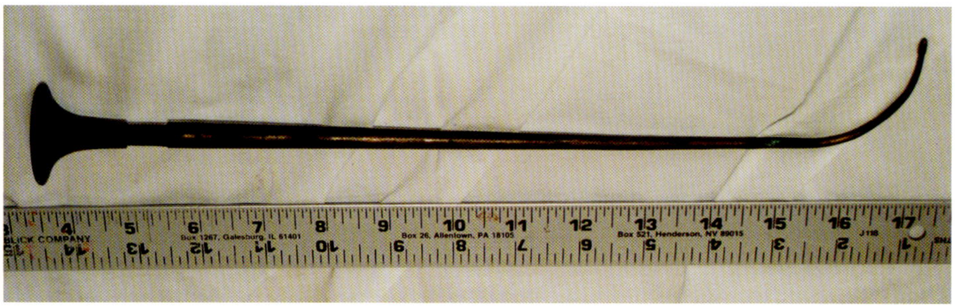

The missing blow pipe from an assayer's kit! By itself, with the mouthpiece, $75.

A cupel tray with a number of unused cupels. This would have been inserted into a furnace for reducing crushed samples of ore. Tray alone, $100 to $200. RB

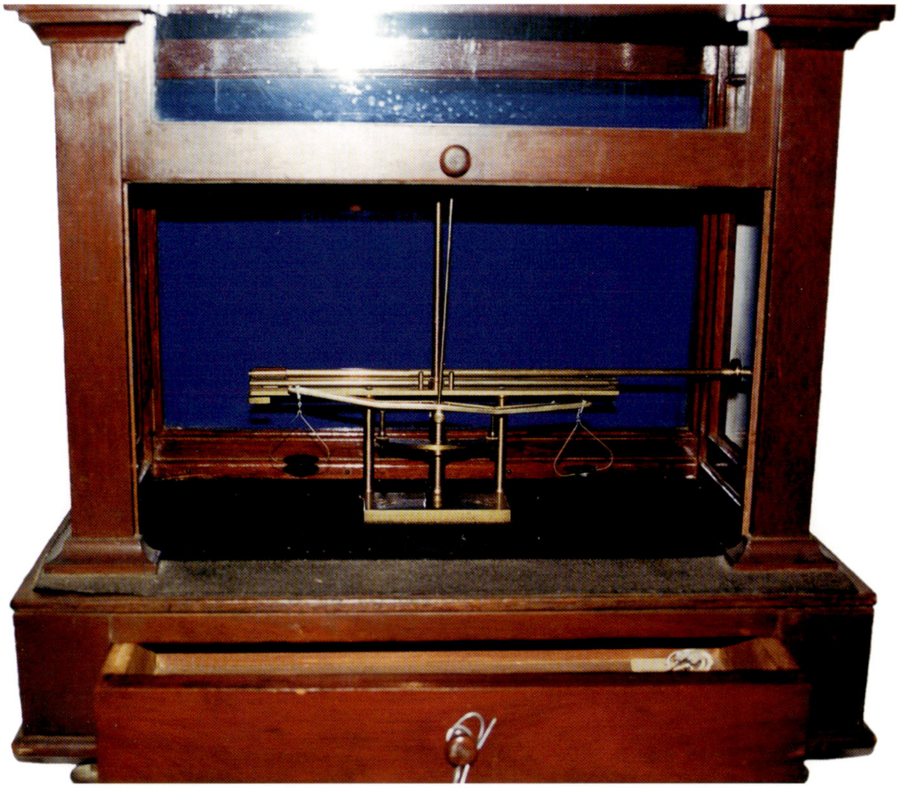

A high-quality assayer's scale. Note the size of the pans and the detail of the case workmanship. $500 to $3,000, depending on condition. RB

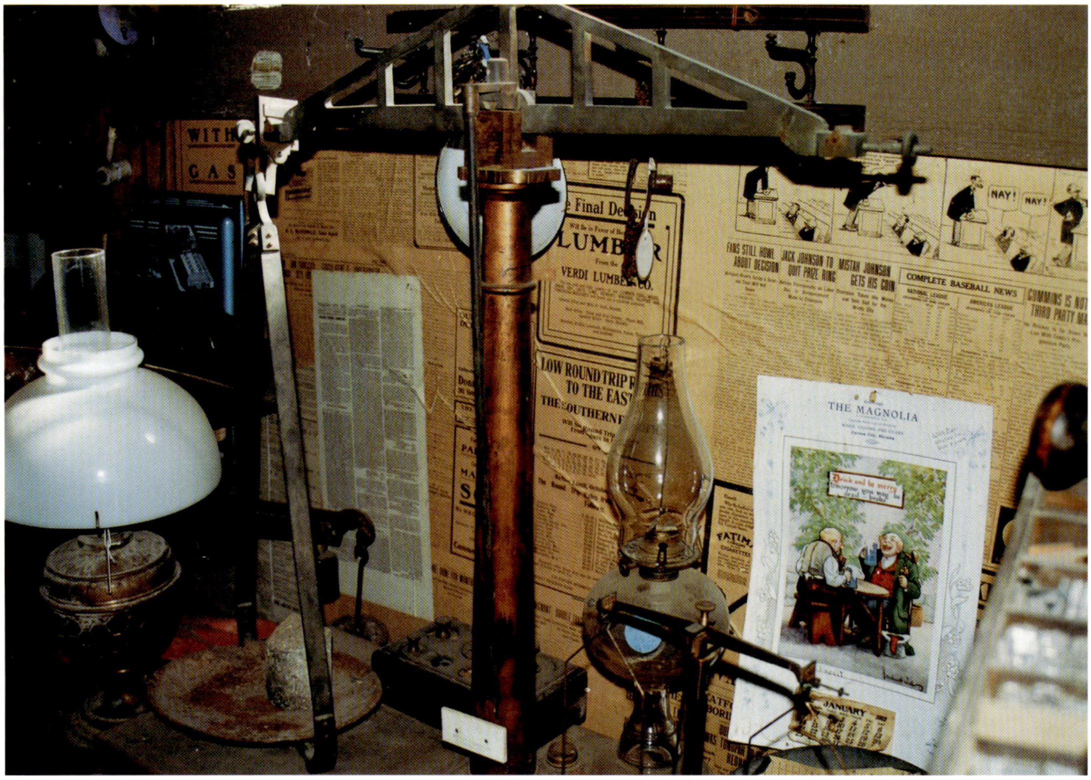

A medium-size bullion scale found in a neighborhood garage. Note the size of the top beam and the cradle on the left side of the scale. This would have been used to weigh small-sized ingots. Relic condition, not complete. In complete condition, $2,500 plus. RB

A large jaw-type crusher for use in an assay house. The small opening to the left would have a rock put into it, then the handle would be lifted up and down until the jaws crushed the ore to small pieces. $300 to $450. RB

A pint-sized flask for keeping mercury in the assay room. $25 RB

An assayer's pulp scale. Not as precise as other scales, but used to give rough measurements. $250 to $450. RB

An exceptionally high-quality microscope for use by a chemist or doctor. This is in fine condition and quite usable today. $1,500 to $2,000. RB

Like new portable assay kit with high-quality scale. Absolutely beautiful. $1,500 to $2,500. RB

The inside of a deserted assayer's shop in a small town in Nevada. RB

Called "The Newkirk" and so marked, this rather heavy item was a patented mechanical gold panning device. It was apparently designed to agitate gold-bearing ore in the pan. It apparently did not work very well, as they are quite rare and few were sold. In like-new condition. $375 to $450. HD

This small assay furnace is known as an "Advance" furnace. They came in different sizes, this being the smallest. This unit stands about eighteen inches high and would have been fired with gas through an opening in the rear. It is complete and functional, although well used. $250 to $375. HD

Above:
This is a small hand-operated grinder for use in assaying. It is about six inches across and solidly cast of iron. Unusual in this size it was made by George White (patented) and came in different sizes. In used but good condition. $150 to $300. HD

Left:
A wall- or bench-mounted cupel maker, this is unusual in its complexity. Made by the Braun Co., it is fully functional and stands about twenty-four inches high. It would have required a stout bench or heavy wall timber to mount to for normal usage. $400 to $750. HD

Right:
Called a "One Stamp," this unit is entirely built of iron castings and would have been used during assaying, or to test production of a specific ore. Very rare today, a few are know to exist. This unit is completely functional. $5,000+. HD

So common they are frequently overlooked, chemical bottles from assaying offices do not bring high prices. This is an old bottle with a correct stopper. $20. HD

A "muller" with its seldom seen heavy iron base. The tool was a common item in assay offices before mechanical crushers. The heavy iron hammer was used to crush ore on the iron base before assaying. Marked "Taylor," $600 to $800. Unmarked $300 to $400.

Perhaps the finest portable assay kit in existence, this boxed unit is about eighteen inches square and eight inches deep. Everything in the kit is compartmentalized and each tool is fitted into velvet-lined trays. The kit contains a folding assay scale that is glass cased. The bottom drawer holds tin-lined boxes for consumable materials, while all the individual tools are beautifully machined metal with ivory fittings. The entire wood box is hard wood (perhaps cherry) and has a leather carrying case to protect it from abuse. $15,000. RB

Chapter Three

BOOKS

In general, there are three separate, distinct areas of collecting books about mining. All three will be in the area of non-fiction. TECHNICAL books are of interest to collectors of mining gear because they detail the progression of the mining industry. In this process they detail and frequently illustrate equipment and techniques for recovering gold or other metals from the earth. HISTORICAL works are of interest for the background they provide the reader in researching and discovering how mining was done and by whom it was written. RECENT REFERENCE books are of much value to the collector, as they illustrate and describe equipment in place and help date the equipment for the collector.

Books are probably the most critical collectible for condition. In the book collecting trade old books are just OLD BOOKS and the value is basically what the buyer and the seller agree upon. Older mining books are a little more than the average collectible—similar to old automobile manuals—because some books have information about a process that has been forgotten or discarded. These books are of interest chiefly to current day prospectors who are still out there looking for the gold in those hills.

RECENT REFERENCE books about mining are usually low-volume printings aimed at a highly specialized collecting fraternity. These books are available if the collector turns over the right rock or makes the correct phone call and they do not seem to lose value, except when they are physically abused or damaged. Some books have been produced in foreign countries—principally South Africa—and are quite rare in the United States. These can be found only by searching through dealers or antiquarian sellers who have an eye toward the mining field of collecting.

HISTORICAL WORKS are frequently biographies or histories of a mining company or mining area. Volumes will be found concerning the history of a particular mine, for instance, and some of these are exceedingly rare and just as valuable. In our companion book *This Was Mining in the West*, we discovered after the book was in print that one extensive set of photos used was part of a promotional book done for the Empire Mine in Grass Valley, California. Quite unknown to either the mine—now owned by the State of California—or us, the volumes of photos were done by the ownership in 1902 to help sell stock in the failing company. How many of the books and photos were initially done is not known, but it is believed that only three now exist. Rare indeed!

So it is not unusual to find local books produced by a company or a historical society that have regional value out of proportion to their value in other parts of the country. During the California gold rush and later the Alaska and Yukon gold rushes, detailed manuals were printed and sold to the unsuspecting greenhorn, supposedly providing him with all the knowledge needed to find his very own gold mine. These books were usually done in an inexpensive format and when found today in decent condition bring rather high prices.

Included in the book section are current periodicals; particularly the quarterlies *EUREKA*! and the *Mining Artifact Collector* are two specifically targeted on mining collectibles. Both are still in print and past copies are a wealth of information and worth trying to locate.

Sutro Tunnel book, $250 and up. Assayer's supply catalog, $100 to $200. RB

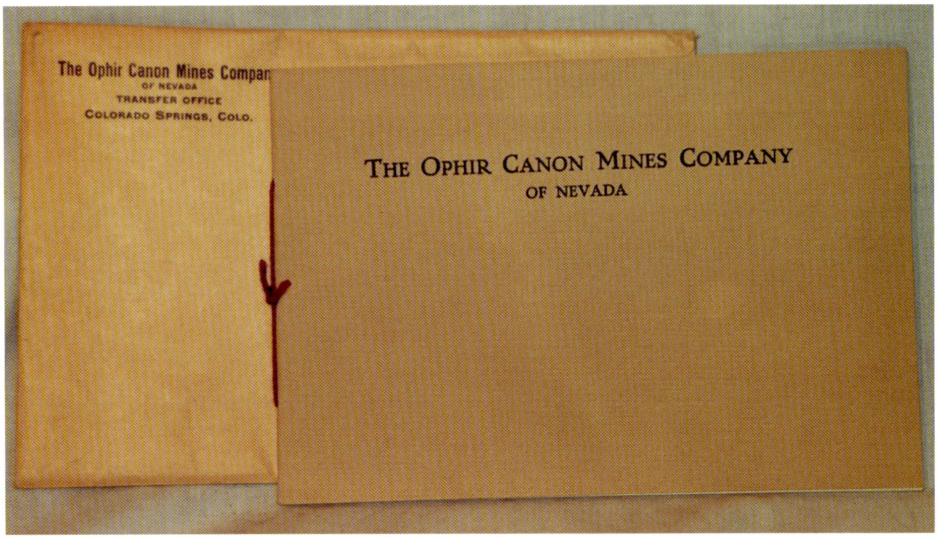

Ophir Canyon Mines Company prospectus with photos in envelope. $200 to $500. RB

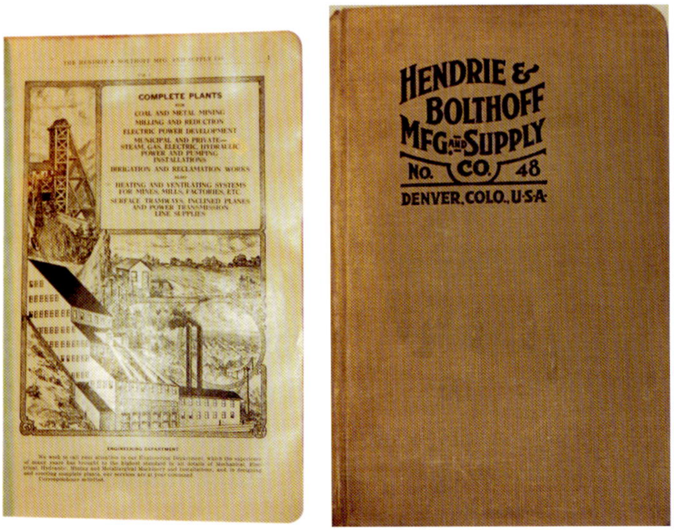

Two mining suppliers' catalogs from different locales. These are quite scarce: $50 to $100 each, in very good condition. RB

Books are included in mining collectibles. Robert Service is way up on the list of miners' friends, actually having spent some time in the Yukon gold fields. His *Songs of a Sourdough* is a good period piece. Depending on condition, $50 to $100 for a first edition. *The Comstock Club* $100 to $125, depending on condition. RB

First edition of *A History of the Comstock Mines* by Dan DeQuille. This book is in fine condition and in a slipcase. DeQuille was an editor and publisher of the *Territorial Enterprise* newspaper in Virginia City, Nevada. The book lacks some correct facts about the mines, but its collectibility is unquestioned. $500 to $700. RB

Pelton Water Wheel Books. 1909 - $100 to $350. The 1922 Catalog is $75-$150. BS

Three miners' books. $40 to $100, depending on condition and age. RB

Leather-covered folio used to record equipment inventories, including photos of the mills and larger equipment. The purpose of the book was to show the mine's assets to prospective investors. Quite rare. $2,000 to $3,000. RB

Nevada Land of Silver in fair condition. Very rare. $300 to $500. RB

The town of Rhyolite, Nevada, is now a ghost town, but was once a thriving mining town with brick buildings and a large population. This booklet was produced to sell the advantages of the town as a place to live and work. It has many photos and is quite rare. $300 to $500. RB

A very good copy of the *Cripple Creek Strike* by Emma Langdon worth several hundred dollars. RB

The Explorers' Miners' and Metallurgists' Companion, undated but probably from the 1870s. $200 to $250. RB

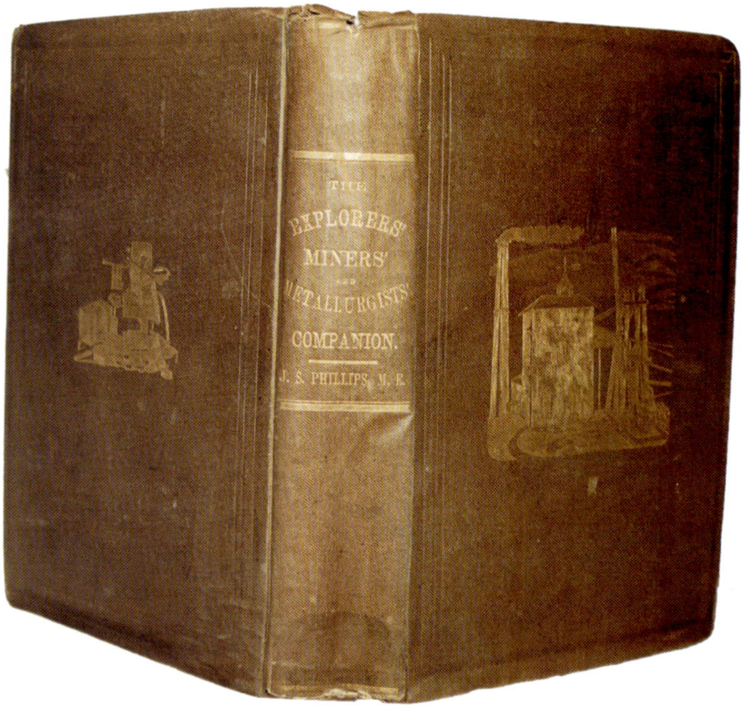

The first three books are technical works dating from the late 1890s to pre-WWI. The fourth is a record of the production of precious metals in the United States dating from 1886. Values on such books vary according to condition and usefulness. $20 to $100 as seen here. RB

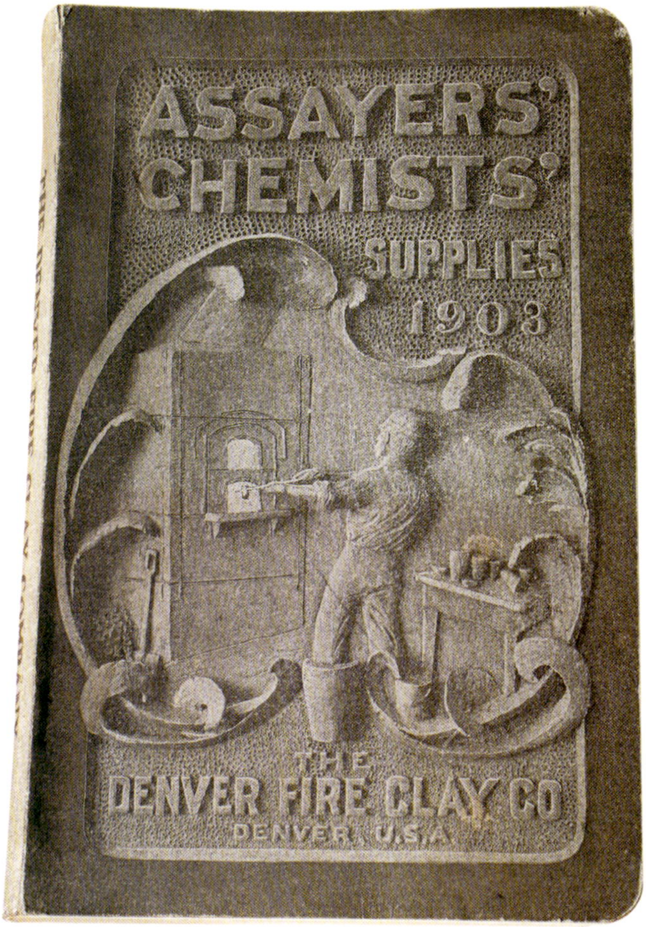

An original catalog from a miners' supply house of 1908. The book is about five by nine inches in size and complete. A very valuable resource for identifying that unknown item in your collection. $125 to $150 and hard to find. RB

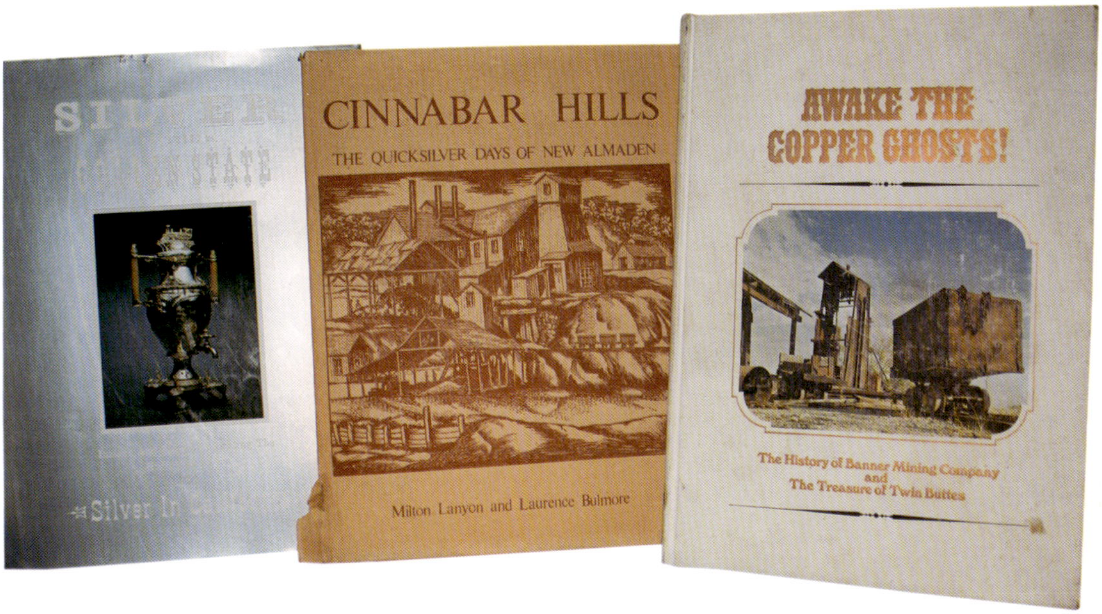

Silver in the Golden State (left) deals with silver produced from the Comstock Lode of Virginia City and Cerro Gordo of California. A limited printing of some years ago, it is worth $75 to $100. The center book is about the New Almaden mines in the California coastal range that produced the lions share of mercury used in mining in the west. It is likewise valued at between $75 and $100. The third book is a "show and tell" about a copper mine in Arizona. Not as rare, it is still unusual. $40 to $60. DP

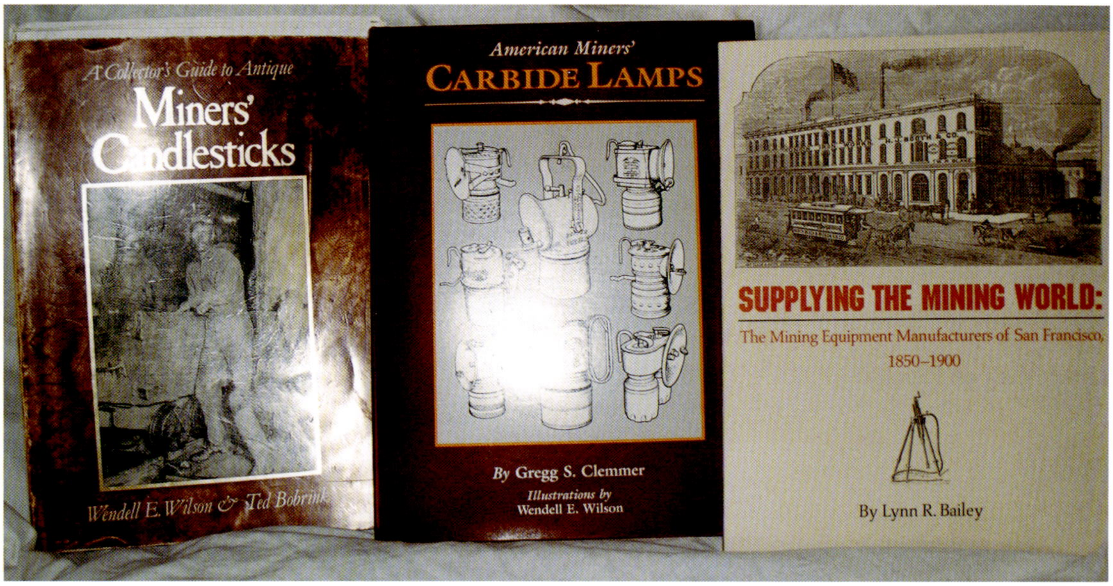

Three books that any collector of mining equipment should have. The two on the right are still in print and available. The miners' candlesticks book is out of print and hard to find, but is the best source of information on candlesticks. Left: $200. Center: $25 to $45. Right: $75 to $125. DP

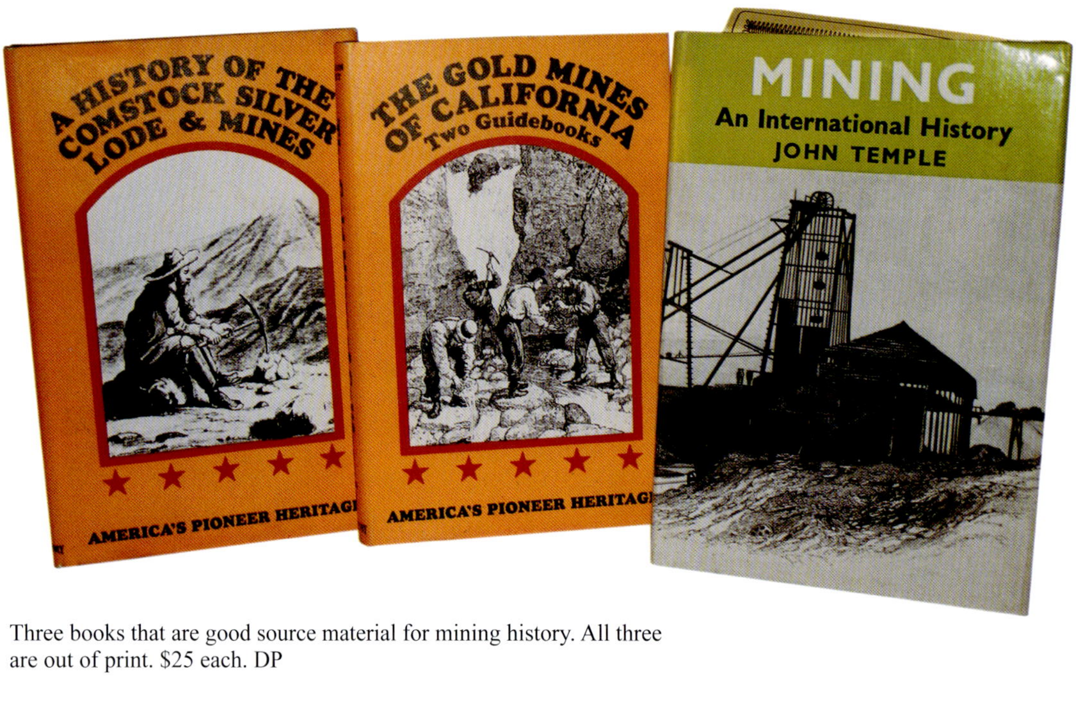

Three books that are good source material for mining history. All three are out of print. $25 each. DP

Chapter Four

CANDLESTICKS

Also known as "stickers" and "sticking tommies," these ubiquitous items were once so common that many collections were started with them. They can still be occasionally found in flea markets, antique stores, and even gun shows. Prices for these plain-jane items have skyrocketed in recent years and it has become an area where fakes have been perpetrated on the unsuspecting collector. Early candlesticks were indeed rough and commonly crudely made, and many of the fakes are made in this likeness. However, early on in the usage of the candlestick, enterprising manufacturers produced them in quantity and many were mass produced in later years using machinery. Early blacksmiths had a great deal of pride in their work—even on inexpensive stickers—and they would try to eliminate any hammer marks or flaws. Bad workmanship on a candlestick is a good red flag that the tool may be faked.

Actually, over a long period of time the basic candlestick holder was refined and patented. It was in fact subject to eighty-eight different granted patents starting in 1872. The first patent issued was to Theodore A. Washburn of Gold Hill, Nevada. Gold Hill is part of the famous Comstock Lode of Virginia City, Nevada. The last of these patents issued was to an individual in Butte, Montana, in December 1917. Some of these probably never saw production and many were no doubt produced and found to be too expensive for the average miner to purchase. Normally candlestick holders were owned by the miner, not the mine, and the miners were even charged for the number of candles they used in the course of their normal occupation, so the candlestick that worked and cost the least was the most commonly used.

Candlestick holders were designed in a variety of ways to make the miner's life easier and safer. The first patent candlestick holder came about as a result of a disastrous fire in a mine in the Comstock Lode of Virginia City, Nevada. The patent was issued for a piece of spring steel that would slide sideways over the flame when it reached the metal and snuff out the candle. In use the candlestick holders were driven into a timber or crack in the rock, or hung from a hook on its side. The candles were then put in the holder portion and lighted. Prior to this, candles were allowed to merely fall out of the bottom of the holder. This unfortunate practice caused the ignition of a dry timber at a lower level of a mine and the ensuing fire killed a large number of miners.

Subsequent patents were issued for all sorts of so-called improvements. Better quality candlestick holders will be found with a folding apparatus and fancy miniature tools welded in the hand loop. We have even observed one or two of these items that were nickel-plated and obviously intended as presentation pieces. Finding one of these presentation pieces is the nirvana of candlestick collecting.

As a sidelight to candlestick collecting, some antique dealers and collectors will occasionally find empty candle boxes. These wooden boxes are nice items to enhance a candlestick collection and depend on condition and age for value. The fact that candles were in common use from the 1860s up into the twentieth century is an indication that they were a reliable source of light and an economical product. The photo of the miners on the cover of *This Was Mining in the West* was taken in 1902 and every miner had a candlestick with a candle in it.

Candlestick holders were later adapted to hold oil lamps and even carbide lamps. By the time carbide lamps were common the soft hat and then the hard hat was coming into use. The hard hat provided a good solid location for the miner's light and also allowed him to use both hands. Eventually, the carbide lamp gave way to the electric light and the open flame light source was out.

Box end from "Crystal Wax Candles," of interest since it came from San Francisco, California. $75 to $200. RB

A candlestick patented by R. A. Miller in September 1897 is both a safety candle holder and oil lamp holder. Extremely rare and may be unique. $3,000 plus. RB

Grant Lamps, so-called because of a famous photo showing President U.S. Grant holding one while on a tour of the Comstock mines in Virginia City, Nevada. The lamps with isinglass windows held a large candle. Unmarked. $750 to $900. RB

A more elaborate Grant Lamp complete with a miner's candlestick that fits into a slot of the top cover. The lamp is accompanied by an invoice for its purchase. $2,000 to $3,000. RB

Right:
Complete boxes in good condition. $300 each. EC

Below:
Stearic Wax Candles box. End and side views of the same box in excellent condition. $500. EC

Slightly poorer condition wax candle boxes (top to bottom), $200 and $250. EC

The best there is. This is the original patent certificate and the candlestick holder that accompanied it. The owner values this at $3,000, but it could be something over $5,000. EC

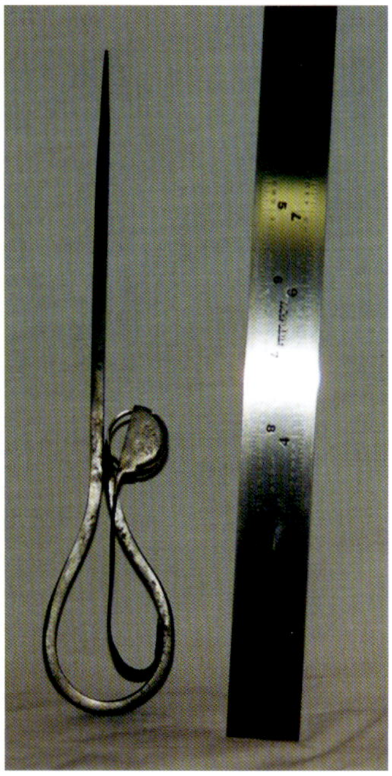

One of the rarest candlestick holders, this is a Washburn patent—one of only five or six known to exist. Price is difficult to place, $4,000 to $6,000 or more. EC

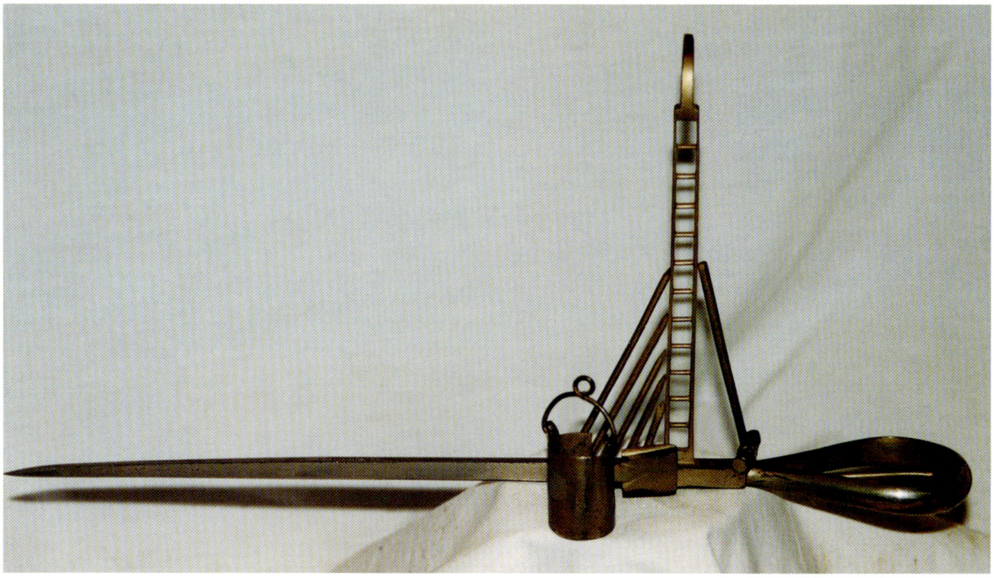

Possibly the finest-made candlestick holder we have seen, this piece was apparently made as a presentation piece and not intended for use. Each part of the holder is a miniature of a tool used in a mine. $5,000. EC

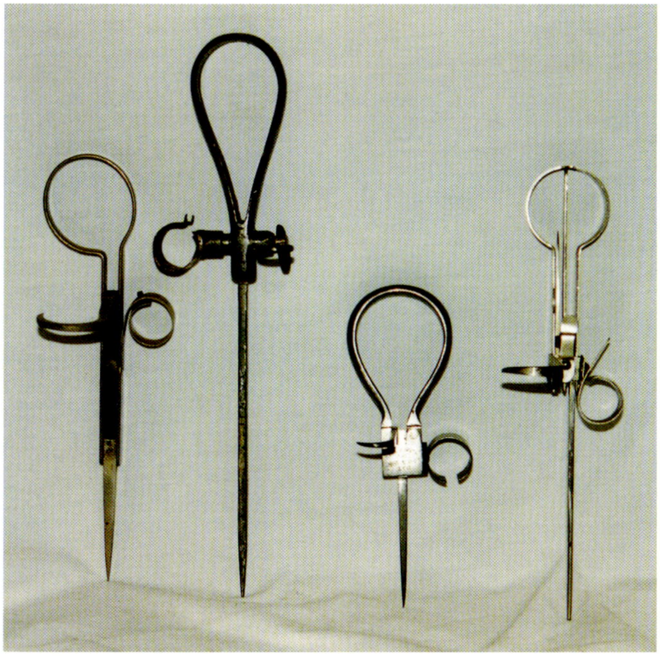

Patented folding candlestick holders were attempts at making the large and dangerous tool more user friendly. They are quite rare these days. From left: a Warbrick, $2,000 to $4,000; a Miners Baby, $3,000 to $5,000; a Hunan Tate, $3,000 to $5,000; and a mechanical folder by Bernier, $3,500. EC

An unusual candlestick from the Nevada State Museum collection. $150 to $225. NM

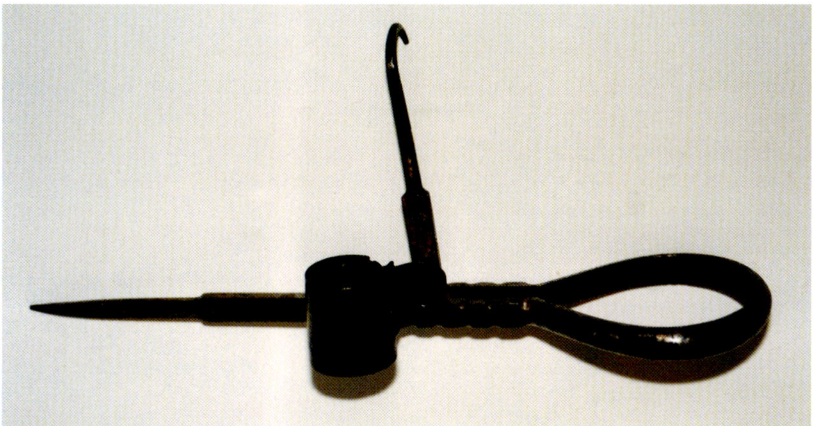

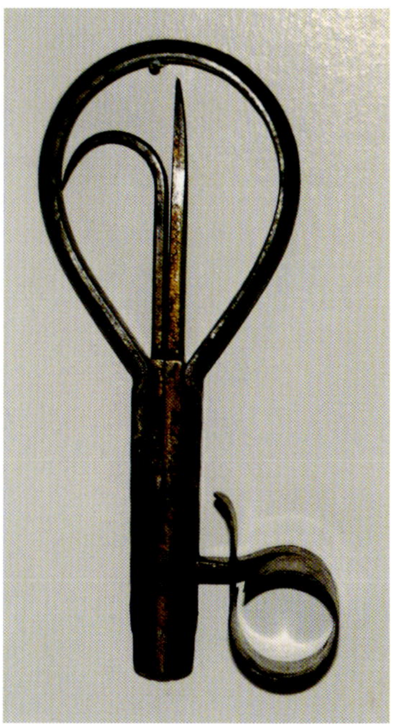

Left and above:
Finely crafted folding candlestick holder by an unknown artisan, probably a Fisk. $2,000 to $3,500. RB

Hanging type candlestick holder, Blacksmith made, claw type of better quality. $175 to $400. RB

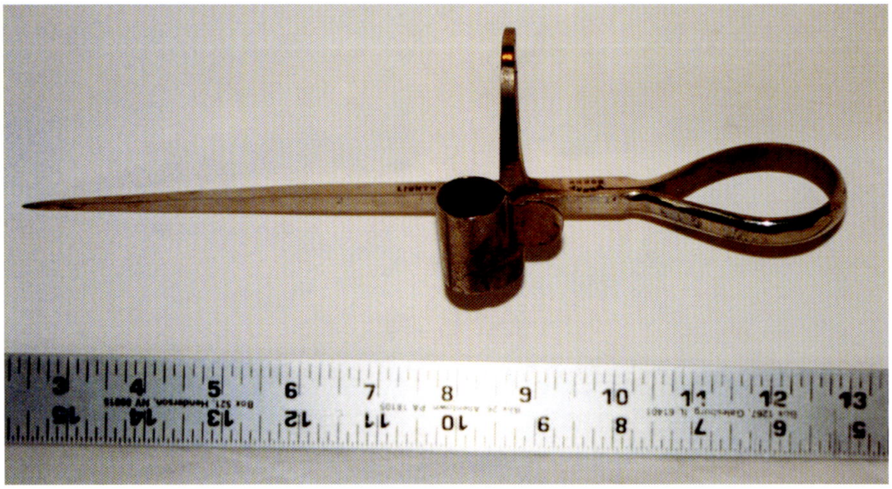

Nickel-plated candlestick holder by Boker, "Lightning" model as new in the box. $450 to $800. RB

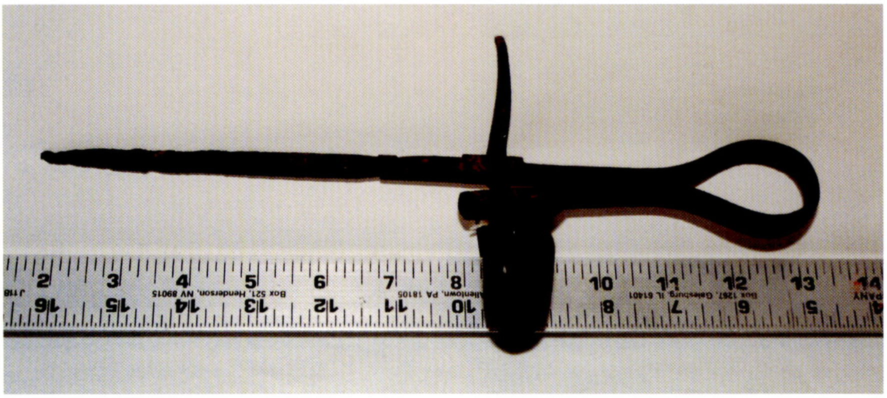

A well-rusted candlestick holder from inside a mine. This is a detachable spike model from an unknown maker. $300 to $500. RB

Unusual candlestick holder designed to be worn on a hat as well as stuck in a timber, probably eastern in origin and fancy. $350 to $500. RB

Wire and sheet metal candlestick holder by Husson. $200 to $350. RB

Husson's miner's lamp or candlestick holder. Very rare. $550 to $800. RB

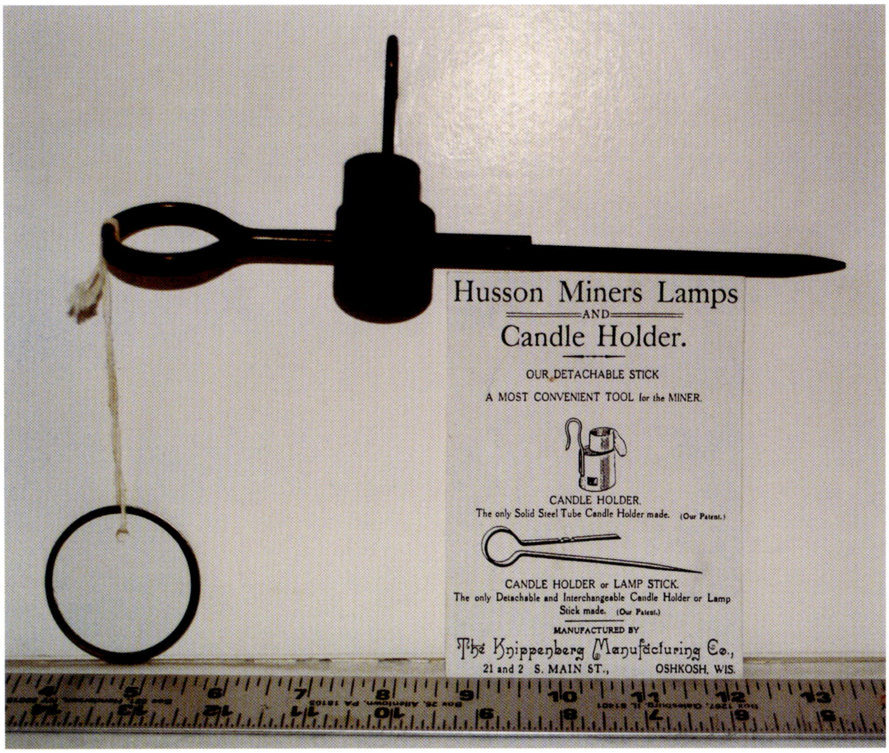

Candlestick holder a blacksmith made to hold an oil lamp. From the ghost town of Rawhide, Nevada. $300 plus. RB

Folder-type candlestick holder by Ramstead and Johnson patented circa 1909. $600 to $800. RB

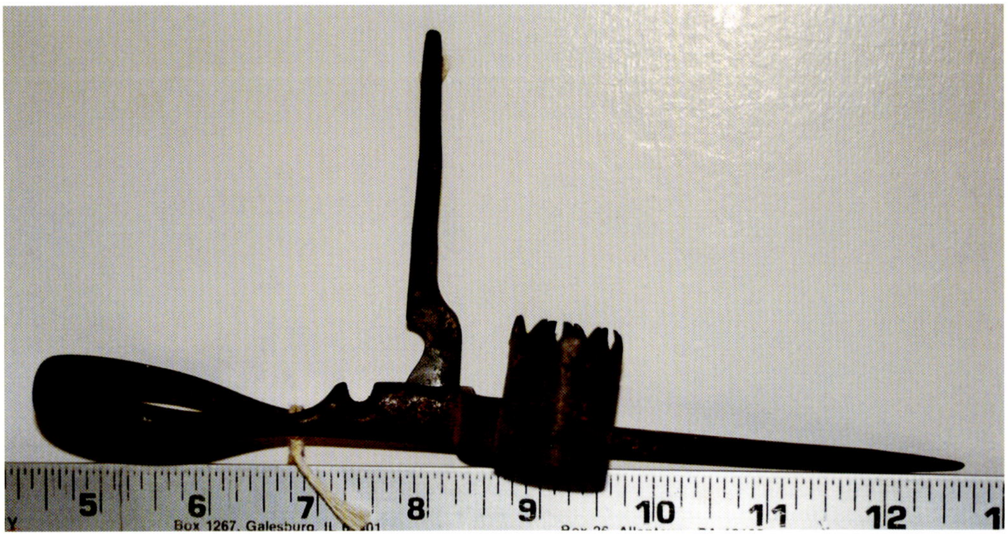

Folder type of candlestick holder that could also be used as a cutter for fuse or other items? Made by Cleaves. $1,000 to $1,400. RB

Very early crudely made folding type of candlestick holder from a Virginia City, Nevada, mine, not patented and locally made. $500 to $700. RB

Ornate candlestick holder by an unknown maker, high quality. $750 to $2,000. RB

Folding type of candlestick holder with the back portion made of brass. Marked "patented" or "patent applied for" but the actual maker is unknown. $2,000 to $2,500. RB

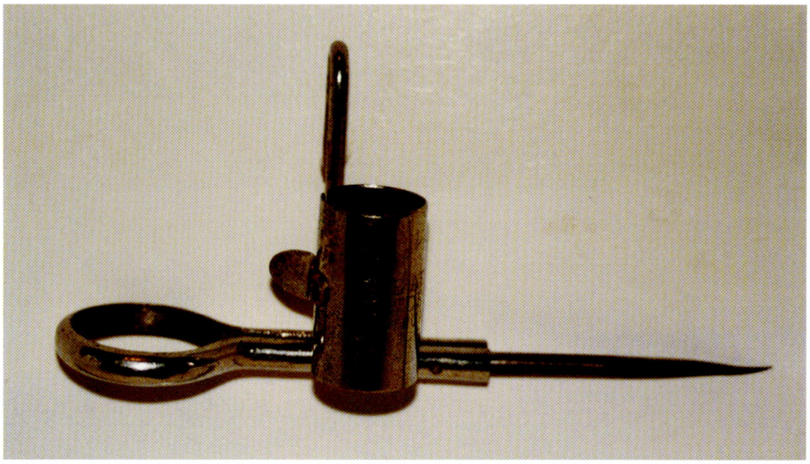

Nickel-plated candlestick holder by the Aetna Powder Company. $3,000 to $5,500. RB

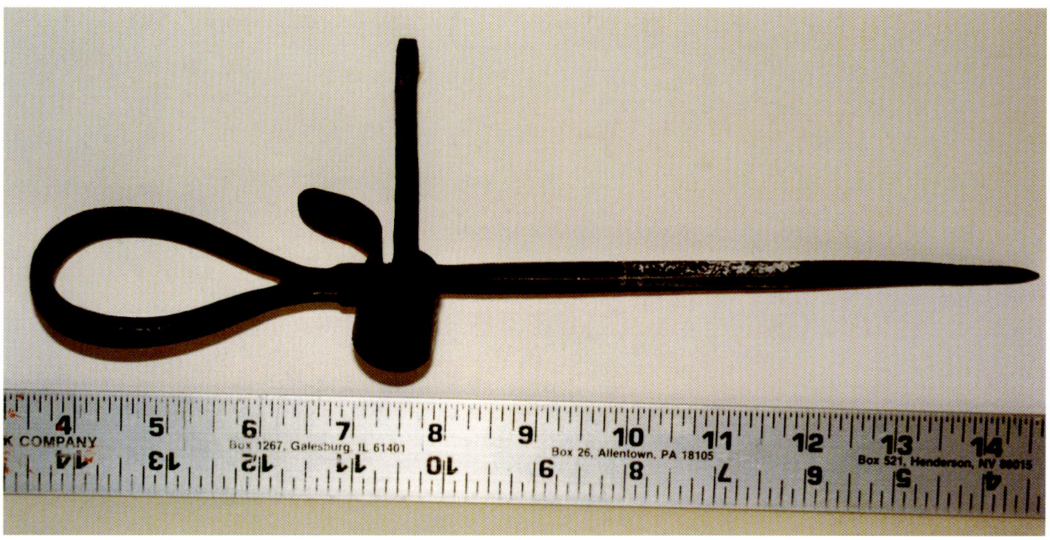

Early simple type of candlestick holder made by the Ludlow Wire Co. with various markings. $150 to $450. RB

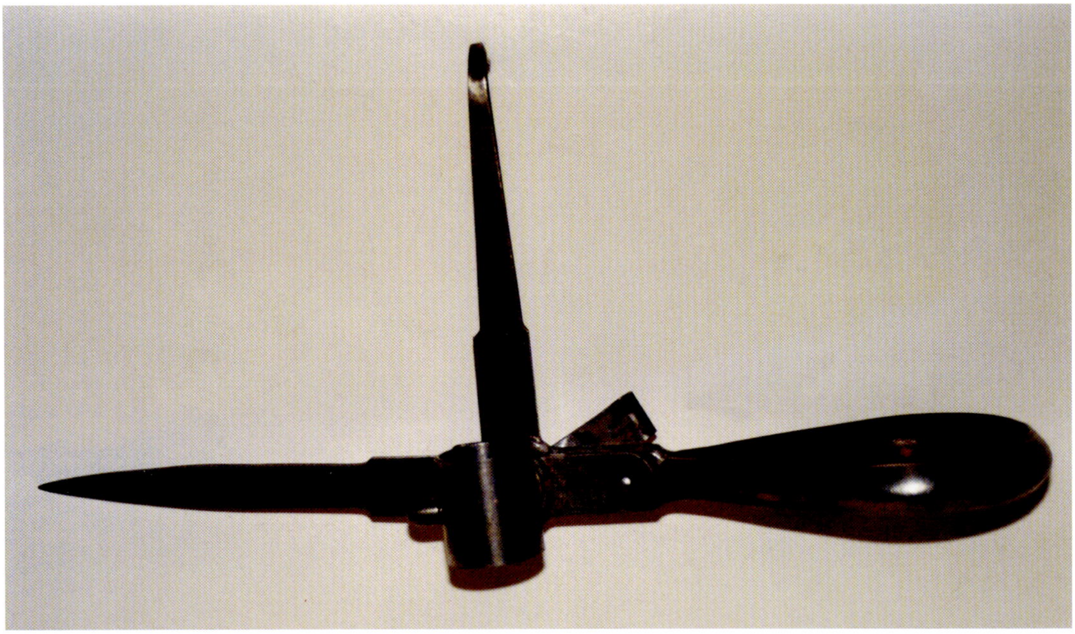

Two views of a patented model folding candlestick holder by Samuel Peterson, Grants Pass, Oregon. Finely made and in excellent condition. Approximately $2,000. RB

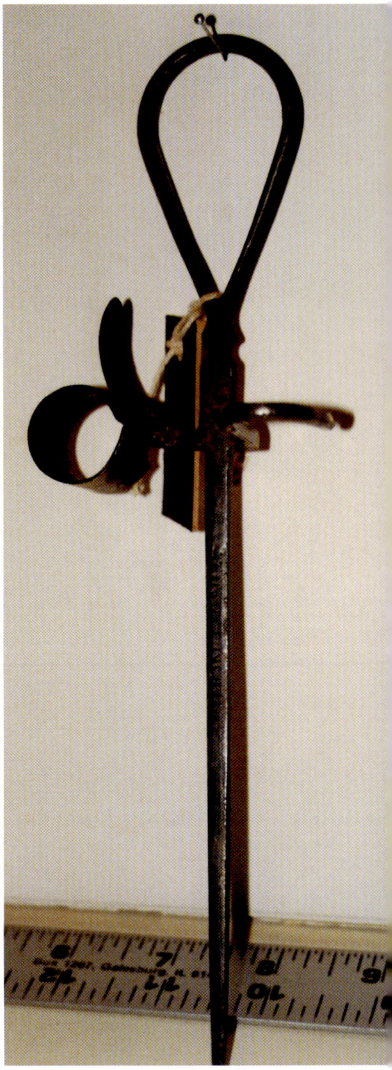

Another hat type of candlestick holder. This one is a bit better made by the Ludlow-Saylor Wire Co. and is known as an Iron Range candlestick holder. $250 to $300. RB

Larger candlestick holder called a spring clip by Edgar Curtis of Baker City, Oregon, in 1904. $1,500 to $2,500. RB

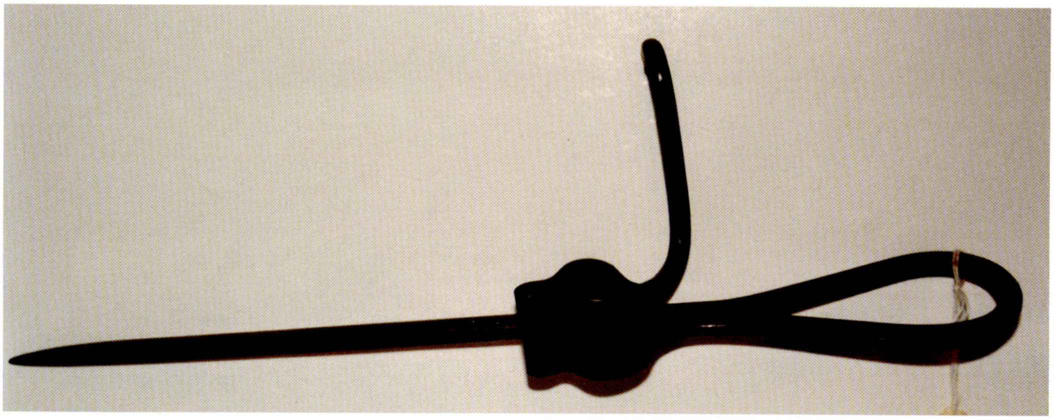

Candlestick holder formed from a single piece of wire by A Varney Co. and fairly common. Varney produced this candlestick holder in nineteen-plus variations or names. $45 to $125. RB

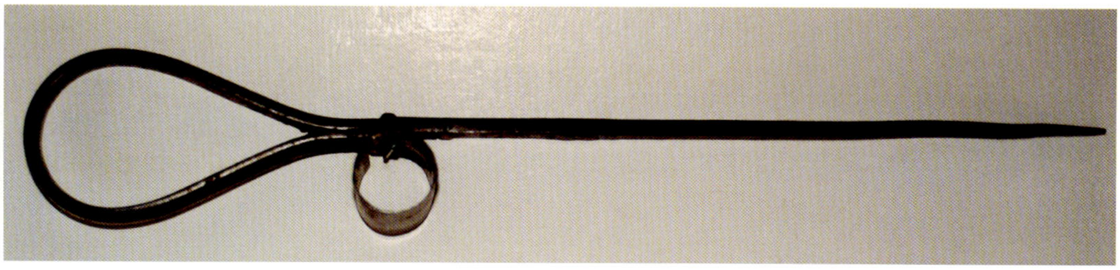

Candlestick holder in its most simple form, known as a Virginia City style, with no hook on the stick and probably blacksmith made. $100 to $150. RB

Very ornate but well-used candlestick holder with a claw. $650 to $2,000. RB

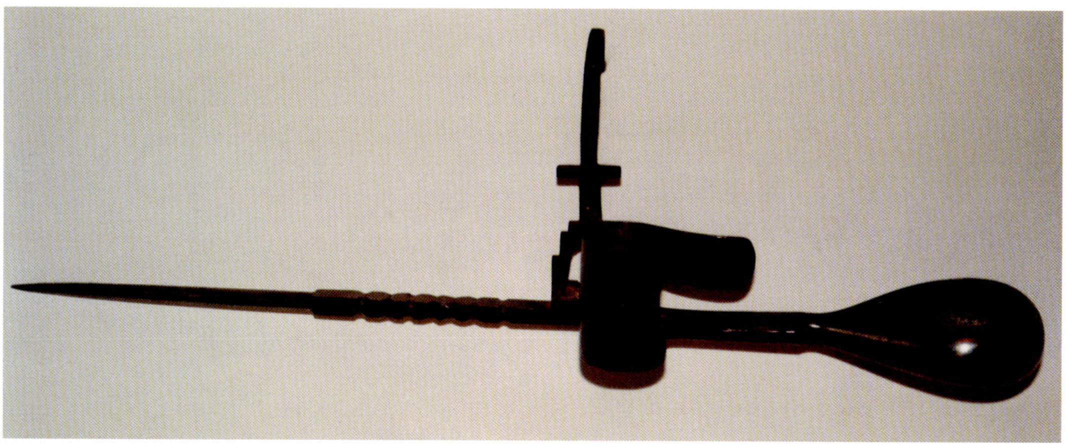

Well-made candlestick holder with a large push-type handle. This type was called a high-grade handle because miners would put wax in the hollow of the handle and stick nuggets into the wax. $2,000 plus. RB

Varney Company produced folder-type made mostly from stampings.
Nickel plated. $5,000 to $10,000. RB

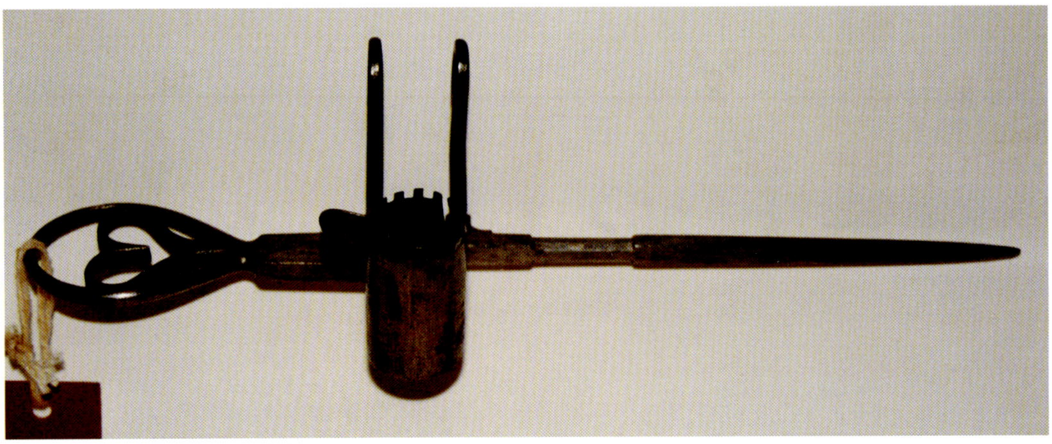

Ornate type of candlestick holder with two hooks for use on a hat.
Midwestern in origin, high quality. $2,500 plus. RB

One of the few known examples of the Washburn safety candlestick holder, patented December 1872.
First patented miner's candlestick holder from Gold Hill, Nevada. $4,000 to $6,000. RB

Known as a "Korean" candlestick holder, this was made in Korea and embellished with silver inlay. They were made for Americans who worked there in early mines. $2,000 to $3,000. RB

An early example of a miner's candlestick holder, maker unknown. $150 to $350. RB

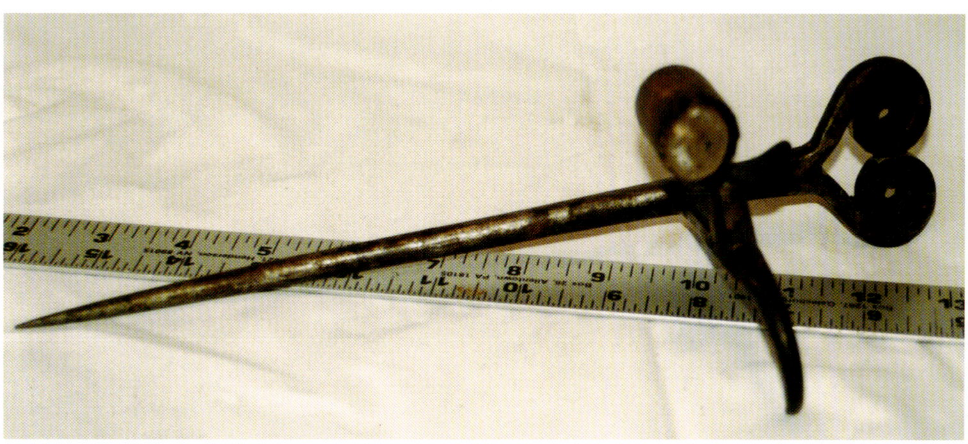

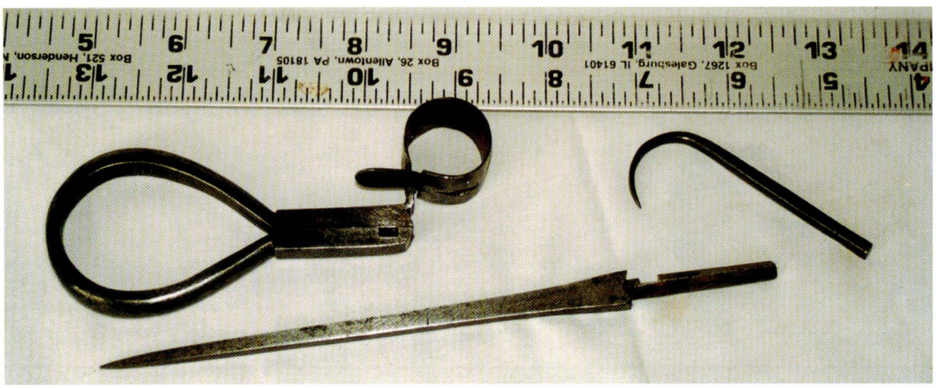

This finely made candlestick holder is designed to come apart for carrying. Workmanship is above average with no markings. $4,000. RB

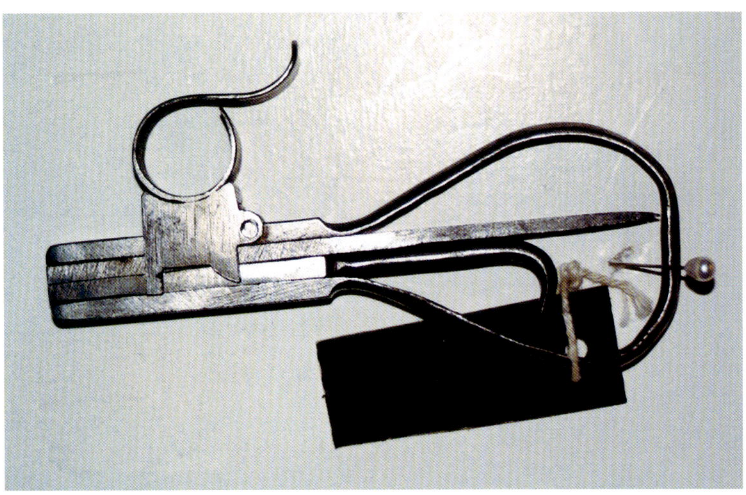

Finely made folding candlestick holder. A Fisk patent from Spokane, Washington, 1909. $3,500 to $5,000. RB

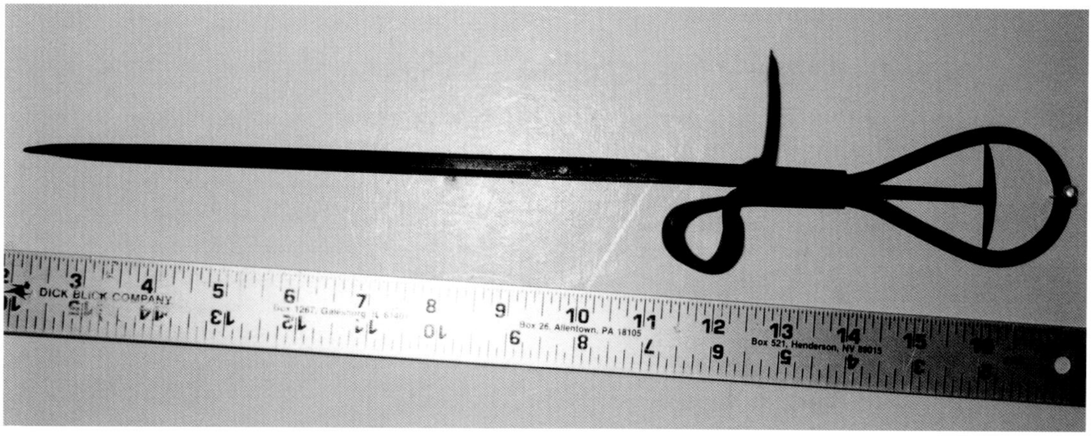

A rather large candlestick holder (about sixteen inches) marked "Placerville" (California). $700 to $1,200. RB

A rather unique candlestick holder where the holder is mounted in a gimbal, allowing the candle to be upright at all times. Patented September 8, 1903. $3,500 to $5,000. RB

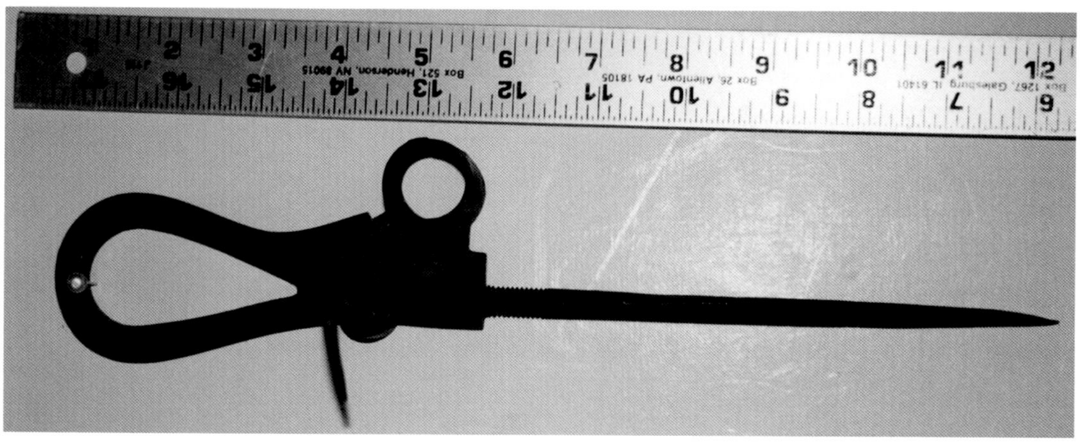

Unique candlestick holder with a heavy base of cast brass and a screw-in steel hook and spike. Not previously known. Found in Goldfield, Nevada, and probably made locally. $2,000+. RB

Folder type of candlestick holder with a vintage candle in place made by Hugh Rice of Fair Play, Colorado, 1881 patent. $1,500 to $2,500. RB

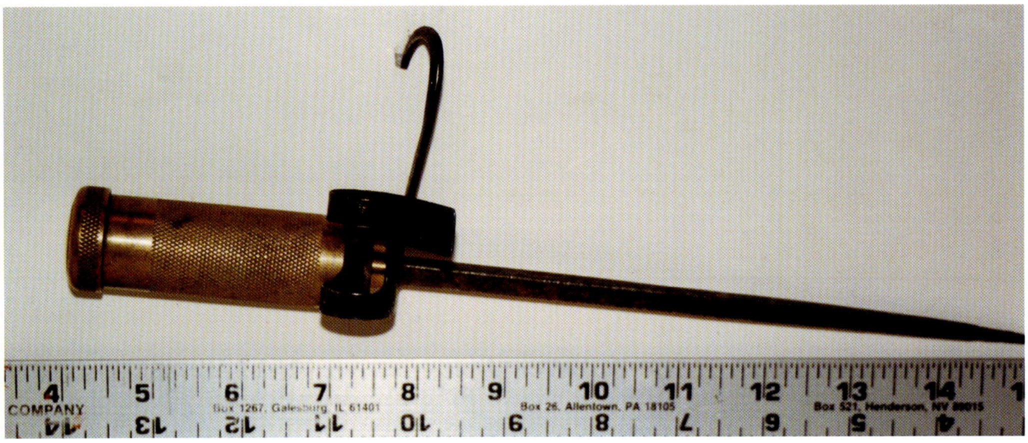

Patented by John A. Lindahl in October 1905, this holder has a brass match safe for the handle. It should be noted that fakes of this candlestick have been observed on the market. $400 to $600. RB

Very fancy tooled miner's candlestick holder. $700+. RB

Hand-forged miner's candlestick with miniature tools in the handle. $1,500 to $2,000. RB

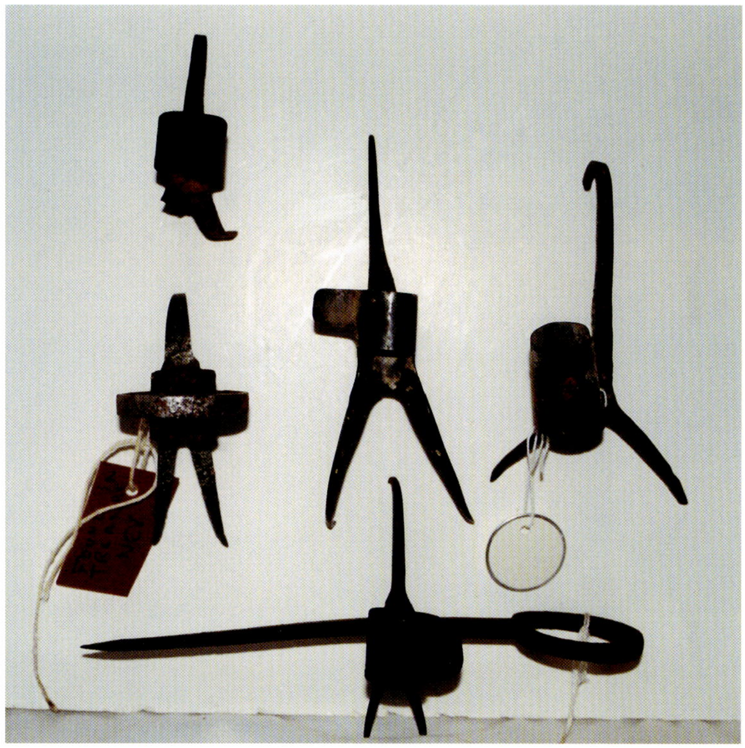

Grouping of miners' candlestick holders known as "claw type" holders. Depending on size and condition, $250 to $450. RB

Miner's candlestick formed from one piece of steel. Probably made by Joseph Kindelain of Leadville, Colorado, about 1906. $450 to $800. RB

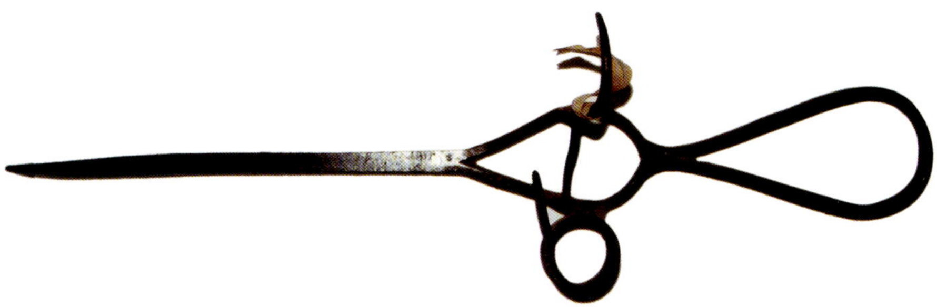

Blacksmith made miner's candlestick from Eureka, Nevada. $450 to $600. RB

Chapter Five

CARBIDE LAMPS

The carbide lamp is perhaps the most common collectible in the mining field. It is still in production in various countries and these cheap foreign-made lamps occasionally show up. The carbide lamp gave a better light than a candle or an oil lamp and was about as reliable. The lamps all depend on a chemical process to produce a flammable gas that is contained inside the lamp. The gas is allowed to escape through an orifice where it is ignited by a spark from a wheel striking a piece of flint similar to a Zippo lighter. The resulting flame is bright blue. The term Carbide comes from the solid form of carbon that is put into the bottom of the lamp. The carbide chemical then has a water drip introduced to it from the reservoir on top of the lamp that creates a chemical reaction, causing acetylene gas to be produced. The gas is then burned to produce light.

The lamps come in all sizes, from small hand-carry units to large room-illuminating globes. They can be dangerous in the hands of an unskilled user as the acetylene produced can be explosive if allowed to accumulate in a confined space. Carbide lamps were in common use virtually everywhere mining took place and the variations can be numbered in the hundreds. The most common lamps can be found for reasonable prices and are a good way to start a collection.

Construction of the lamps was commonly of brass, as it was easy to work and more resistant than steel to the corrosive chemicals in the lower portions of the lamps. Some lamps can be found made of aluminum and in reasonable condition these will bring a slightly higher price. There is an old adage in antique collecting that says, "if you don't have it all you don't have anything." This is especially true of carbide lamps.

The collector is admonished to avoid partial lamps or lamps in very bad condition. The only exception is for extremely rare specimens, when ownership is the first consideration.

In our travels to ghost towns and mining camps of the west we have been fortunate to see many large rusting metal cans with screw top lids. These cans were bulk supplies of carbide for use by miners in their lamps. To uncover a can in pristine condition would be a rare find indeed. When exposed to air the carbide chemical takes on water from the atmosphere and turns to a white powder. This powder is corrosive and should be avoided if possible. Carbide is difficult to find today but is still manufactured.

The finest collection of carbide lamps perhaps in the world. EC

Extra containers for carrying carbide into the mines were fairly common at one time. However, they have slowly vanished from the market and are now considered rare. They are also frequently mistaken for ink holders! All of these extra containers have tops and are in fine condition. From left to right: $250, $500 (an Eveready), $500, $300, $250, $300, $500, and $500. EC

Just as a miner needed extra carbide to finish a shift they needed the other ingredient, water. While often mistaken for carbide containers, these are in fact extra water containers. From left to right: the Justrite is $750 and the large Stanklin is $200, while the Greer is $400. EC

These three flasks are actually combination carbide and water flasks used for replenishing the miners' carbide lamps while in the mine. From left: $500, $250, and $250. EC

Carbide was produced and supplied by a number of different manufacturers. The two larger containers are in good but not perfect condition—showing some rust on the tops, which is common—but with bright colors on the sides. Left to right: $50, $25, and $300. EC

Four more carbide flasks: the two in the center are from Auto-Lite and Guy's Dropper, two makers of commonly used carbide lights. From left: $200, $300, $200, and $1,000. EC

Carbide was also available for a time in shaped flasks directly from the manufacturer. These tins are the size of tobacco tins of the same time period and are rare today; note the difference in top sizes. From left: $150, $100, $125, $85, and $60. EC

Three more extra flasks: the flanking two are carbide or water flasks, while the middle one is an oil flask for oil lamps. $75. EC

Flasks for carrying either water or oil into the mine. $50 to $100. EC

Large carbide lamps were not common, but are not especially valuable. Left to right: $150 and $300. EC

This is a nice way to display a collection of carbide lamps. We think this was originally intended for holding cups. EC

One of the better known makers of carbide lamps was the Justrite Company. Repair kits came in two different sizes and were usually handled by stores or mine commissaries. Left: $200 to $250; larger, shown open, $300 to $400. EC

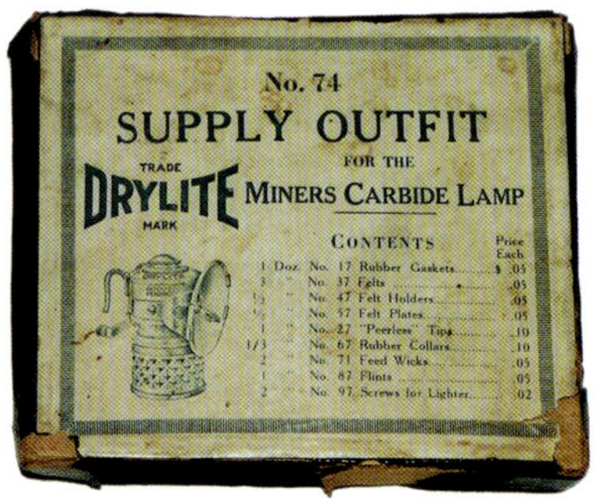

Repair kits were also produced by other lamp manufacturers and these two are from the Hansen Company and the Drylite Company. The Hansen kit is quite rare, $1,000. The Drylite kit is more common. $200 to $300. EC

Finding new carbide lamps is not too common and these are later production. Blue, $75. Red, $65. EC

More carbide lamps new in the box. Auto-Lite lamps, $100 and $150 (left to right). The Justrite supplies in boxes are unusual enough to command higher prices, $100 and $275 (left to right). EC

Carbide lamps were sold and used for other purposes, as this wooden box example shows. The Brilliant Search Light Mfg. Co. produced this example that hung on the belt and supplied the gas to the burner and reflector on the hat of the user by means of a rubber hose. Quite frankly, we would not want to be close by if the hose came off and started spewing flammable gas around in a dark place. $150 to $200, if the lamp is all there. EC

Not all carbide lamps are common and these are among the rarer lamps. Condition on these is very good, but all have been used ("fired" in collectors' vocabulary). From left to right: American, $3,000 or more; Millburn, $2,500; Maumee Duplex, $3,500; and Eveready, $3,500. EC

More high-value carbides, from left: Pathfinder, $1,500; a Union Carbide lamp, $3,500; G B Pushbutton, $1,500; Rex, $1,500; and Zar, $3,000. EC

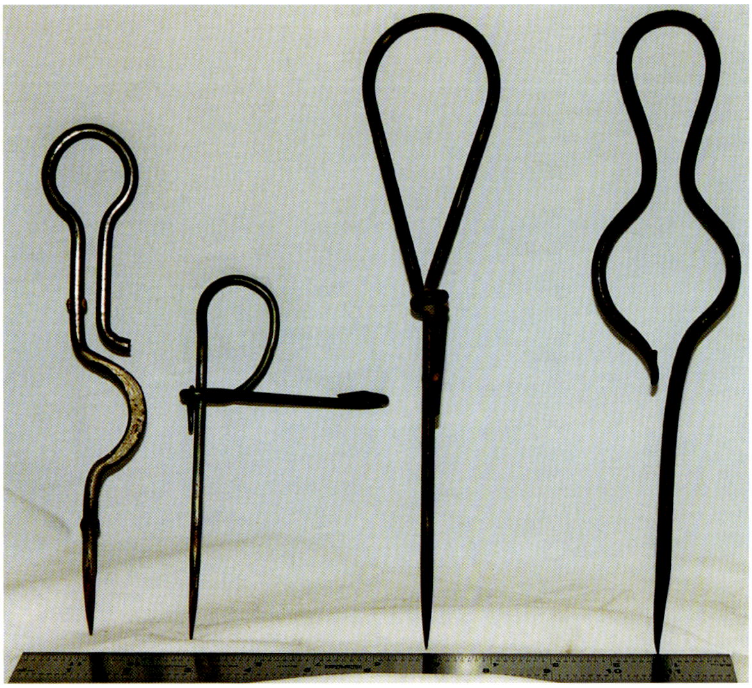

These unusual stick holders are actually used to hold carbide lamps in place of candles and were designed to hold particular lamps. $75 to $200 each. EC

This carbide lamp/burner was apparently designed to light up a large area. No reflector, and it may not have had one to begin with. $125. EC

A large carbide lamp known as a "Buckeye," $150 to $175. EC

Two more large-size carbide lamps. These would have been considered eight-hour lamps. From left: Millburn, $400; and unnamed, $250. EC

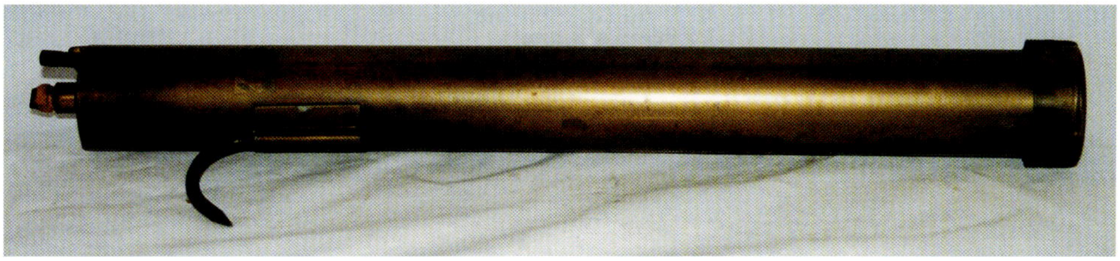

The Arnold long carbide lamp was designed to be hung from a timber. $2,500 to $3,500. EC

Each of these carbides was designed to be attached to a hat or helmet and did not have handles. From left: Defender in Nickel plate, $500; the Anton square top, $2,500; Hansen cap lamp, $1,500; and Simmons, $3,000. EC

Carbide lamps, all with handles and hooks for carrying or hanging. From left: Brite Lite, $2,000; Guys Dropper, $125; Baldwin, $125; and Justrite, $175 to $225. EC

Larger carbides were made and used in mines and these three are good examples. From left to right: $200, $300, and $200. EC

Another large carbide, this one by Wolf Co. $300. EC

Large carbide lamps called an Uncle Sam and the Baldwin. The Uncle Sam, $350; Baldwin, $350. EC

Large Carbides. $100 each. EC

Carbides were made of different materials and these two are of aluminum. Left, $200; and right, $175. EC

Four different carbides, from left: two by Full Moon—the brass lamp, $300 and the nickel lamp, $500; and two lamps by Baldwin—left, $500 and right $300. EC

Four different carbides made by the Wolf Co. from left: $300, $500, $700, and $250. EC

Three fine carbide lamps with sticker holders from left: Baldwin, $150 with the sticker; an Anaconda, $500; and JustRite, $150. EC

Three large carbides referred to as four-hour types. $80 to $100. EC

Two DuRLite lamps flanking a Wolf. Note the slight differences in the DuRlite lamps. $80 to $100 each. EC

Two look-alikes: the Little Giant on the left, $400; and a Copper Queen, $4,500. EC

Very nice looking carbides of smaller size from left: ITP, $250; Justrite, $50 to $75; Victor, $150; and Arrow, $350. EC

Three more Wolf carbides. The two at the ends, $70; the brass, $125. EC

Mid-sized carbide lamps or four-hour types. From left: brass ITP, $100; nickeled, $135; Scranto, $350; and a Pathfinder four-hour, $1,100. EC

Carbide was the lifeblood of miners' lamps. The mines bought the carbide in large tins of about twenty-five pounds. This tin is in pristine condition, $125 or more. SB

From left: large carbide by Guys Dropper, $1,000; a crude-looking Wolf, $150; and a later Wolf, $100. EC

Chapter Six

CLOTHING AND ACCESSORIES

Clothing is a difficult area of collecting because of the real possibility of fakes. Hats and accessories are probably the most common items for collectors to find and these can be dated from period photos to an approximate era. Early miners wore all sorts of head gear, as is evident in *This Was Mining in the West*. Hats with metal or fiber attachments that held oil lamps and later carbide lamps are a much later development. Fiber molded "hard hats" with metal appendages for holding carbide or electric lamps are fairly common and little different from those in use in mines today.

It is common to find parts of old shoes in mining camps and occasionally one can find boots or parts of boots that were in use by miners. Sadly, most are found completely worn out, as they were used hard. Early miners in California and Nevada wore the famous Levi denim clothing and as most collectors know, early examples of these items bring premium prices. Later oilskin-type clothing and rubber boots can occasionally be found but are difficult to authenticate.

The watchword of clothing collecting is *caveat emptor* (let the buyer beware)!

One shoe bottom (left) with an attachment meant to raise the wearer off the surface of possibly corrosive liquid. Rather rare, $125 to $175. RB

A very early miner's hard hat. This one is made of leather formed into shape with a lamp holder on the front. $65 to $150. RB

California gold rush belt buckles from the 1840s to 1850s. There are over twenty different types priced from $300 to $3,000. RB

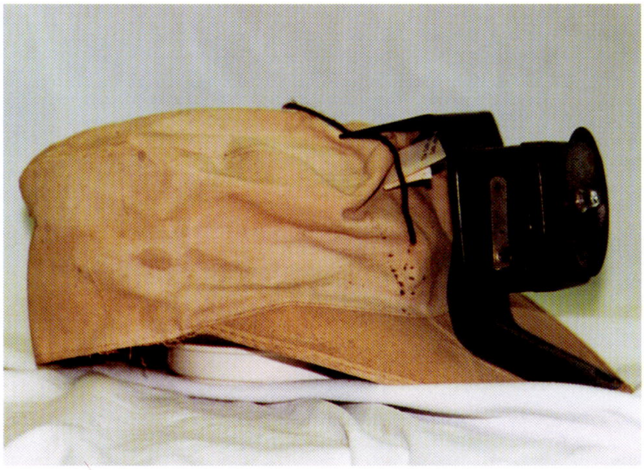

One of the earliest examples of a cloth miner's hat with an electric light attached to it. Rare. $125 to $175. RB

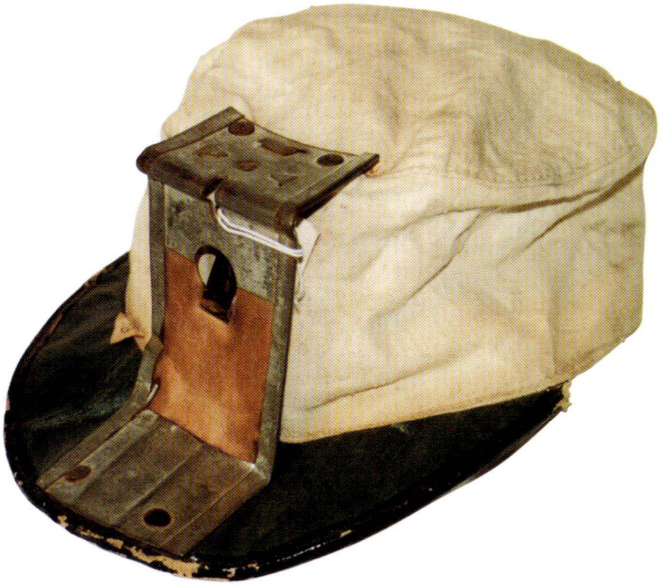

Miner's cloth hat with an oil cloth brim designed to hold either a carbide or oil lamp. $75 to $90. RB

Miners' shoes or boots were subject to hard usage and they sometimes adapted them to fit their own needs. This pair of shoes has leather uppers attached to wooden bottoms that are cleated with steel edges for better wear points. RB

Chapter Seven

EXPLOSIVE GEAR

Cap Tins

"Caps" in mining terminology does not refer to a hat but rather to explosive caps. Early explosives were in the form of black powder charges that were usually put up in small paper tubes for ease of transportation. Occasionally one can find metal kegs that contained black powder used in the mines. These are rare and can be well dated by the manufacturer on the outside of the can. The most common size for early kegs is the twenty-five pound keg, as these were easier to transport and store. Black powder is graded by size and blasting powder is the largest grain size, but not the only one used in mines. Other sizes were made by the powder mills, but these were primarily for use in firearms of the time. Rarely, tins that have a sliding cover and look like oil cans are found that are labeled for powder. These were used to carry bulk quantities of black blasting powder into the mine.

This is all by way of introduction to caps and cap tins. Caps were highly explosive devices still in use by miners and made using fulminate of mercury and other compounds—later a less volatile chemical. If found they should be considered extremely dangerous and avoided. They could be easily set off by a sharp blow or exposure to heat or electricity. They were transported in well marked wooden boxes that held small metal tins of the caps. The caps were taken out only as needed and carried to the working area of the mine in wooden or metal boxes with wooden inserts to avoid inadvertent ignition. A single cap is strong enough to blow off a man's fingers. There were many manufacturers of blasting caps and collecting the small tins and associated items is difficult today, but can also be very rewarding.

Powder Boxes

We are going to combine a couple of categories of items into one area. Boxes that held explosives were always made of wood to avoid the possibility of static charge buildup and a subsequent explosion. The boxes can still be found in mining areas and some of them can be found whole. Collectors frequently only keep the ends of the boxes, which were painted or embossed with the manufacturer, usually including a date. Full-size boxes are rather large and do not display easily, so only the marked ends or sides are commonly kept. Since black powder and later dynamite deteriorated with age dating was important, as it allowed the powder man to rotate his stock. Dynamite boxes should be approached with caution, as old dynamite has the annoying tendency to leach out the nitroglycerin used in the sticks. If a box is found out in the open and it appears greasy or slick, avoid it! Losing a hand or your life is not worth the price of the box.

Other powder boxes will be found that were used for carrying a specified number of charges into the mine. These boxes are usually compartmentalized and are not very large, and always made of wood. In use, the powder charges or dynamite sticks were placed in holes in the wooden boxes and carried to the working face, where the drillers would have prepared a series of holes. The powder men would then put the charges into the holes and fuse them using the detonator caps and fuse. The fuses would then be lighted in a particular sequence to blow out a section of rock. The powder men then moved a ways off to be safe and waited until the charges exploded one by one, until all the charges went off. To forget the number of charges placed was to invite death or injury. Typically, powder boxes and cap tins were carried in separate containers by separate people to avoid an accidental discharge.

Detonators

Detonators come in a variety of sizes and most depend on a mechanical system to generate an electrical charge. They are in fact miniature dynamos. The charge is transported through wires to the caps, which are inserted into the dynamite charges. When the handles of the detonators are depressed or turned sharply the charge is generated and sent to the cap. While these are not particularly common they can be found, usually in good condition. These were expensive items necessary to miners and later replaced the fuse and match method of detonating an explosive charge.

They were made by a variety of manufacturers, typically by the sellers of powder or caps. They will be found with brass tags and are very utilitarian in nature, but are also finely crafted, as they were intended for hard use. There are also battery-powered detonators that are smaller and finely made, but were less common. The military used smaller versions during wartime and these occasionally turn up. A variation in type is called the pull-up type because the electrical charge was created by pulling up on the handle rather than pushing it down. This type will have a pronounced plate on the bottom where the user could put his feet for leverage.

One final caution about collecting explosive gear: BE CAREFUL!

Small metal-hinged box with wooden insert for carrying blasting caps. $75 to $100. RB

Another detonator, this one a pull-up type made by DuPont. DuPont is perhaps the best known of the powder and explosive equipment suppliers. Note the large steel plate on the bottom, allowing the user to leverage with his feet while he pulled sharply up to generate the electric charge. $500 to 700. SB

A common type of electrical detonator, this one by Laflin and Rand, a famous maker and supplier of explosives. The patent date is 1878. $750 to $1,200. SB

This strange looking device is a safety fuse crimping tool. They were made by DuPont and by the California Cap Co. The DuPont is more common at $300, while the California tool is $500. EC

This DuPont detonator is of the more conventional push type and is in average condition. $350 to $700. SB

Fuses for explosives are still manufactured in one form or another. These two brands of fuse are complete unopened rolls. $50 to $75 each. SB

Wooden carriers for both powder and caps were the norm in mines and they were typically locally made. These two carriers—the smaller on the left for caps and the larger on the right for powder (dynamite)—are ingeniously crafted and contain no metal parts. From left: $65 and $100. SB

A "Victor No 1" Detonator. This dynamo provided electric current by turning the crank. Wires connected to the terminals carried the charge to electric blasting caps. $400 to $500. SB

Fuse cutter. $500. DK

Blasting cap tins are one of those items that seem to get tossed by everyone. Blasting caps were packed in small quantities and disbursed as needed. The tins are now remarkably valuable and highly collectible. These examples are all in like new condition and are quite rare. From top left, clockwise: $75, $35, $75, $45, $100, $75, $250, $50, $200, $50, $200, and $250. SB

Circuit testers are ohmmeters used to test wiring circuits before blasting. Relatively common. $75 to $200 each. SB

While the push or pull type detonators are more romantic, later battery-type igniters are more common. $125 to $200 each. The dynamite sticks in the foreground are real wax covers that have had the explosive removed and placed on wooden rolls for display. About $10 each. SB

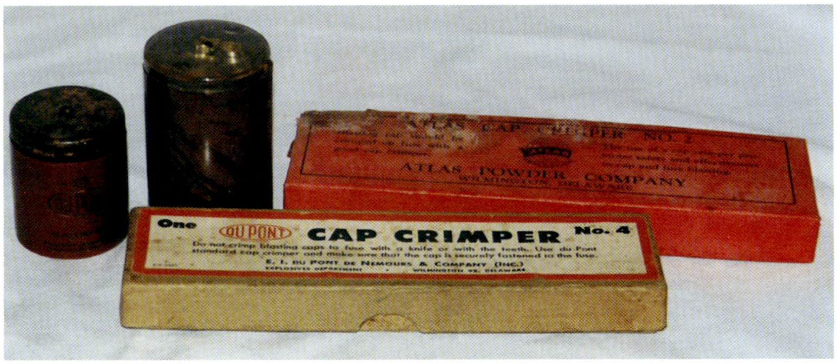

Cap crimpers are simple tools used for crimping fuse inside a cap. New in the box like these are beginning to generate some interest. The DuPont tool in the foreground, $50; the Atlas tool, $75. SB

Early blasting was done with black powder in the mines. It was carried to the working face of the mine in cans as shown. This can could hold about eight pounds of powder and had a spout cover attached with a chain to avoid loss. The powder was poured into the cover as a measure, which was then put in loose or wrapped in paper and inserted into the drill holes. It was an effective but messy process and dangerous to use. $350. RB

Five different powder igniters, all twentieth century items. $125 to $250. EC

Cap tins, from left, top row: California Cap, $40 each; Atlas tins, $50 each. Bottom row: in boxes, $350; $40, $40, and $40. EC

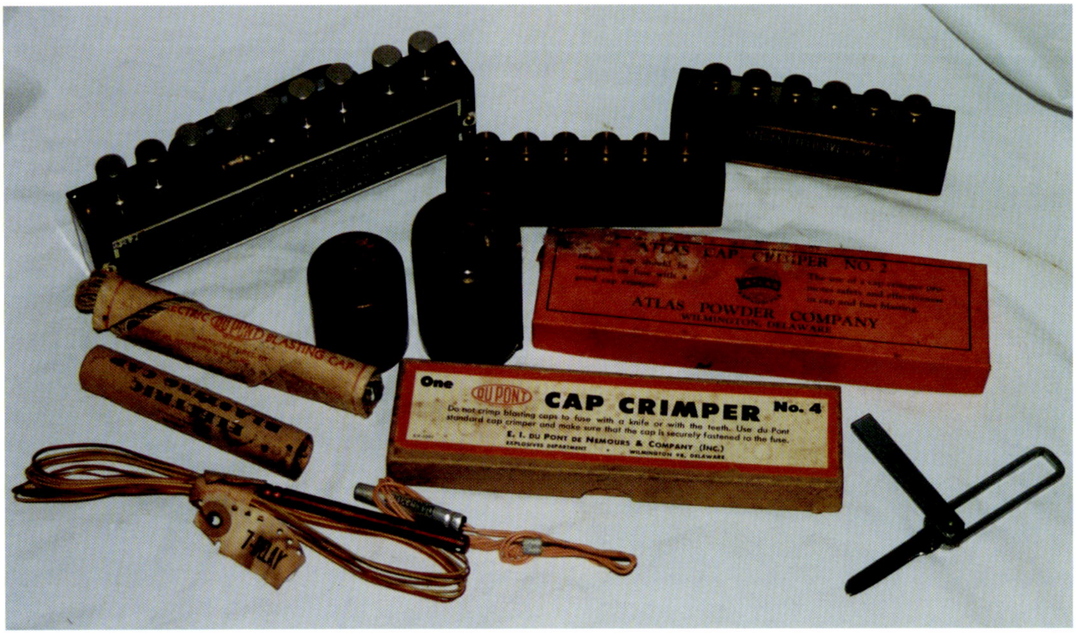

A display of explosive-oriented items from the Smith Brothers collection. Entire display, $200 to $300. SB

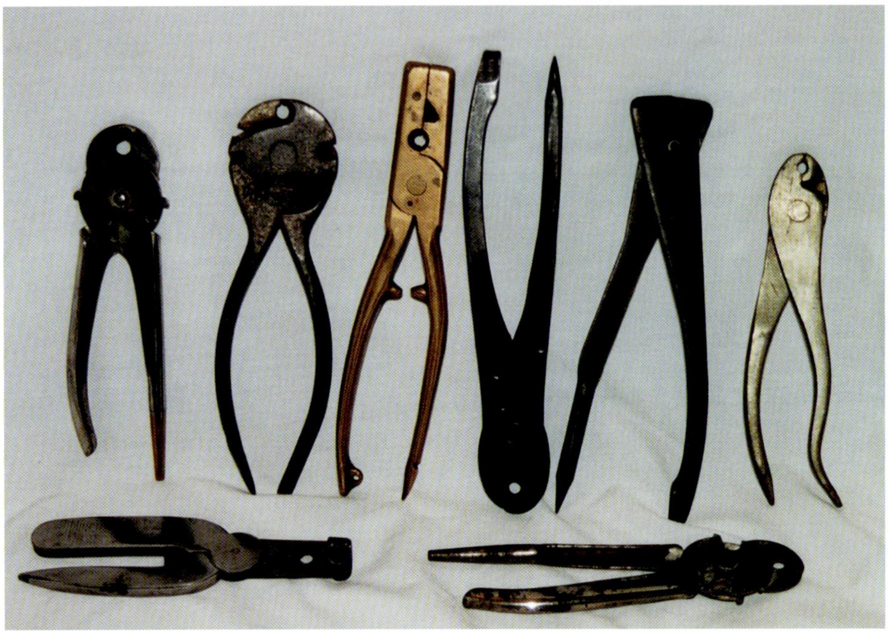

A good sampling of fuse crimpers all by different makers. Clockwise from top left: $100, $75, $300, $60, $50, $65, $400, and $100. EC

Blasting paper was explosive impregnated heavy paper used as a tube or wrapper for loose powder in drill holes. Unusual, $75 to $100. RB

Powder boxes came in different sizes and these are ten-pounders. The more common types are twenty-five pounds. Condition is very good on all four and they are priced accordingly. About $100 to $150 each. EC

Explosive detonator in good condition. $450 to $700. RB

A nicely named (Samuel Nash, Georgetown, Colorado) powder can for carrying black blasting powder into a mine. In fine condition. $350. RP

This photo was taken to illustrate the brass plate on the side of a DuPont detonator. Widely used, these blasting machines will bring premium prices in this condition. Compare to an oval brass nameplate shown elsewhere. $500 to $900. RP

This item could be classified under either explosives or advertising. This glass-covered cased display board appears to be factory original and is in fine condition. The small card in the upper right corner reads, "This case contains no explosive material." The date at top center is July 1911, and the assumption is that it was made then. Probably more exist, but none have been for sale that we could find. $5,000.

Chapter Eight

FRATERNAL AND POLITICAL

One of the earliest bastions of union organizing was in the mines. Unions of one sort or other developed early in the history of mining and were always a contentious subject. In addition to unions, fraternal orders were very much in vogue from the earliest times. Secret societies were common in the gold rush days of the California mother lode diggings. Societies such as the Masons and the Knights of Pythias were common and today, many of the better preserved buildings in such districts are or were lodge halls.

To show their loyalty and membership, members purchased or were given badges with elaborately embroidered ribbons attached that they wore to meetings and during parades. These ribbons are found today with metal tags attached and occasionally with members' names engraved on the metal portions. They are frequently very colorful and highly collectible. Less common are coins or tokens that a member was expected to show on demand to other members. The most collectible items will come from famous mining camps that are now ghost towns or completely destroyed, such as Bodie, California, and Aurora, Nevada. Regionality is probably the most important consideration in this area of collecting. For whatever reason or reasons, some western ribbons will bring inordinately more money than equally old items from eastern or midwestern mining regions.

This medal is from Cripple Creek, Colorado, dated 1901, and is of solid copper. No indication what it was for. Because of condition, $75 to $150. RB

Denver, Colorado, 1914, was where this Elks pin was given out with a cast miner's candlestick on the top. $300 to $450. RB

Two ribbon banners from Elks fraternal lodge events in Manhattan and Tonopah. The two mining camps were within a day's auto ride on the roads of November, 1911. $75 to $100 each. NM

Fraternal orders were a big deal in the mining camps of the west (and the east) and were initially the closest thing to a union. Allegiance was usually shown by wearing pins with an attached banner. Some of these were especially ornate, with the highest values assigned to those from early mining camps. If the mining camps were short lived, so much the better. These four banners are all from unions, but those from the masons and other societies are equally interesting. $200 to $350. RB

Left:
Extremely rare miner's union ribbon worn for parades, etc., from Bodie, California. Items from Bodie are scarce. Up to $2,000. RB

Right:
Two miner's union ribbons. Very fancy, these ribbons were given to miners to proclaim their membership in a time when unions were sometimes a fractious subject. Tuscarora ribbon, $200 to approximately $900. RB

Chapter Nine

GREASE LAMPS

The earliest "grease lamp" was usually little more than a small piece of metal about the size of a deep saucer. The miner would fill the depression with grease of one sort or other, then soak a length of rope in it and light the end of the rope. The result was a flame that beat the total darkness of a mine but was frankly not as good as a candle. This very primitive type of illumination was used early on, but quickly fell out of favor as candles became available or were made by someone in a mining community. The lamps could be made out of almost anything—including rock and animal horn—as long as it would hold a suitable quantity of grease. Examples are seldom found today and real specimens are hard to find.

Later grease lamps were made using hooks and reservoirs with a wick, but the principle of operation is the same. In some areas these were known as "froggers" and were popular in the eastern mines. Alladin's lamp is a good example of an early grease lamp. Some early grease lamps have been found in the west in old mining town basements, but were apparently disliked and seldom used. They were in common use in European mines and a lively trade exists in some areas, but values are not generally very high.

A very early grease lamp, this variation could use two crossed lengths of rope soaked in oil to produce four flames at once. Blacksmith made, $85. RP

Left:
A very early and crude miner's grease lamp. The wick was held in place by the latch on the top of the pan. Also known as a frog or frogger lamp and European in origin. $200 to $300. RB

Right:
Another grease lamp of the "frogger" type, this one is European in origin and factory made. About $100. RP

Chapter Ten

GOLD SCREENS

Gold screens show up in the most unusual places. These manufactured products have been in production and use to the present day. They are commonly found in places where the owner has no idea of their use. This writer recently bought one from an antique dealer who swore she had used it to strain milk when she was a kid. Having no reason to doubt her, we bought it and then explained to her how they were supposed to be used. The brass screens are built using differently sized screens to separate larger flakes and nuggets from fine powder-sized gold. The screens stack or nest one inside another with a large pan on the bottom. In use, the gold is put into the top, where it is shaken with the largest screen at the top and the smallest at the bottom. The gold sifts down through the gradually reducing size of screen. The screens are then disassembled one at a time until all the gold is recovered and the jewelry-sized pieces are taken off first.

The screens were or are made in different sizes, ranging from about six inches across to upwards of twelve or fourteen inches across. A complete set would stand several feet high, but in practice a stack of six or seven screens of different openings would be normal. Condition is very important, as near new sets occasionally turn up. In our experience the screens are usually made of high quality brass. Occasionally screens made of tin are found, but values are not as high as brass screens.

Gold screens are still in use by miners today, particularly by placer miners. The collected gold is poured into the topmost screen and agitated. The gold sifts down through progressively smaller screen sizes, with powder collected in the bottom basin. These are complete sets with tops and bottoms. Construction is almost always of brass and individual screens are always labeled as to the size of the screen. Larger set, $250 to $400; smaller, $150 to $300. SB

Right:
A set of eight gold screens from the Denver Fire Clay Co., these are unusual because they are made of tin, not brass, and each screen has its own bottom tray. These are about nine inches in diameter. Not expensive, these were recently purchased for $240. HD

Chapter Eleven

HAND TOOLS

Crimpers

These normal-looking hand tools appear to all the world like a set of dog nail trimmers. In use they were for crimping the metal outside a blasting cap down onto the fuse inserted into the open end of the cap. Manufactured by many suppliers, they were and are so common as to be overlooked by many collectors. They are in fact one of the more collectible of mining tools and equipment. They have been made of steel and brass or copper and will almost always be found well used. Virtually every supplier of caps or powder sold them under their own brand name, even if they were made by a subcontractor.

Picks

The miner's pick is commonly found and should not be confused with a prospector's pick, which is smaller and designed to be used with one hand. The miner's pick is usually a full-sized pick in heft but just as commonly is found with a shorter handle to facilitate the confines of the interior of a mine. The pick used by a laborer today is much larger than a miner's pick. There have been a few patented picks, including folding models and models with replaceable points. They were all designed to do one thing: dig rock out of a mine. Early picks may also have a blunt and a pointed end. This style allowed the pick to be used as a hammer as well as a digging tool.

The prospector's pick is still found in hardware stores today and unless of verifiable age is of little value.

Shovels

Like picks and other common implements, shovels are not items that bring much value as collectibles. Shovels did go through a series of evolutions, starting with hand-made specimens with riveted blades that could be replaced to the forge-formed shovels we know today. Specimens can still be dug up today (without handles of course), but the value is in the eye of the beholder.

Pans

Gold pans were in use from the earliest of times and there are some identifying characteristics that should be looked for, with size being one of the first. Early pans were nothing more than steel dinner plates that could be used to recover gold from a stream bottom. Development of larger pans with steeper sides and bottoms came about as a result of the search for more gold. Finish will always be well used, with the early pans having no paint or chemical finish.

Photos exist that show large pans in use and the best advice we can give is to study those photos carefully to help authenticate an "old gold pan." Early gold pans can be made with several sections forming the sides. These were then tinned or soldered together to form a watertight pan with a flat bottom that would catch gold in the angle.

A steel ingot stamp for marking bullion ingots from a particular mine. This one is from the Yellow Jacket mine of Virginia City, Nevada. Quite rare and seldom worn out or tossed out, they exist today because of collectors' interest. $1,500 to $2,500. NM

A grouping of ingot stamps from mines and assayers. These are very rare. $800 to $2,500 each. RB

The original seal for embossing wax seals from the US Mint in Carson City, Nevada. One of a kind, we think. $3,000+. RB

A true artifact that came out of a deep mine in the Comstock. This is a good example of a vintage miner's pick. $60 to $80. RB

A fancy miner's pick. Probably used by a shift boss or supervisor. $300. RB

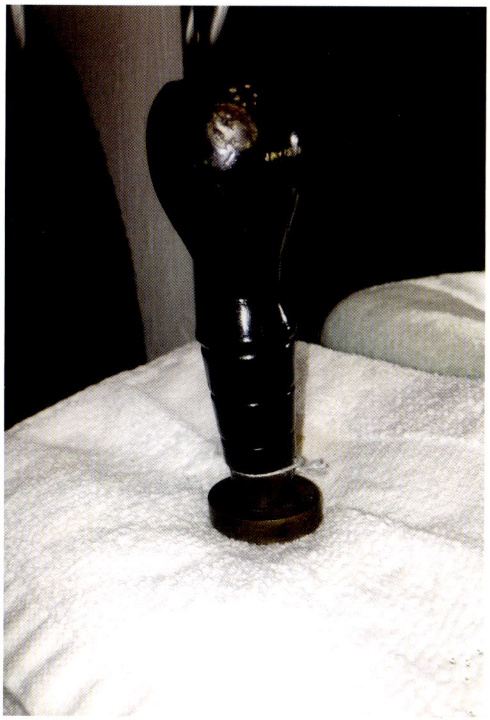

Wax seal for Candelaria Mining Company, Candelaria, Nevada. $750. DK

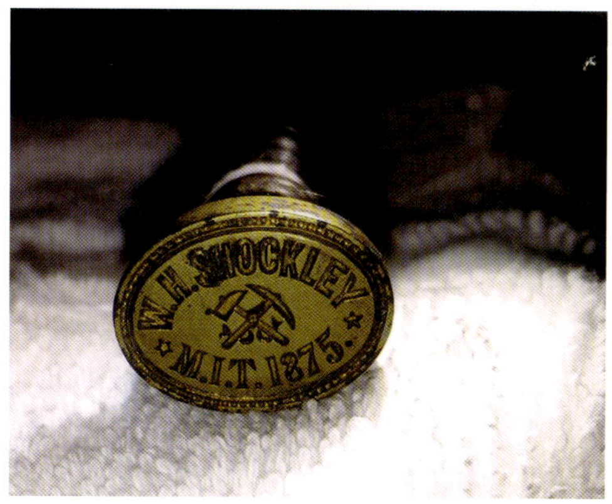

Wax seal for W. H. Shockley 1875, Ely, Nevada. $500. DK

Wax seal for American Mining Company, 1863. $500. DK

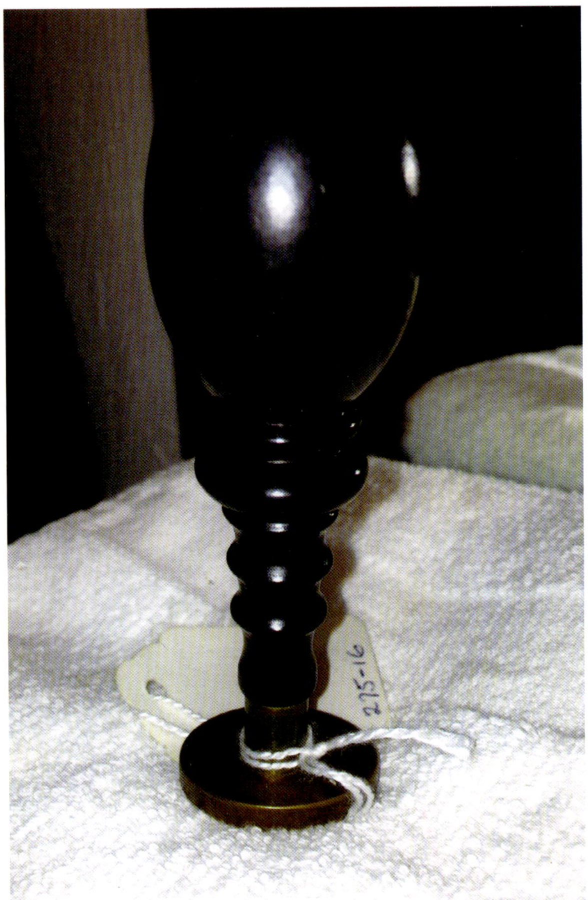

Wax seal for the Rawhide Queen Mines Company, Rawhide, Nevada. $500. DK

A "spoon" for cleaning out drill holes prior to charging with explosive. $75 to $100. RB

This odd looking item is a copper or brass spoon used to clean out dust or muck in drill holes prior to setting a charge. $90 to $125. RB

A metal flask for holding lubricating oil for drilling machines in mines. $45 to $125, depending on marking and manufacture. RB

Chapter Twelve

HEAVY EQUIPMENT

Crushers

For purposes of this book we will include stamp mills in this area. Crushers by definition are larger pieces of equipment designed to reduce ore into small pieces, either by simply smashing the metal-bearing rock between two weights or by grinding pieces into dust. The earliest crusher is the ARASTRA type, which is simply a large heavy rock dragged repeatedly over ore until the ore is smashed into fine particles small enough to pass over a sluice box for recovery. In practice, the arastra was of a size large enough to be pulled by a horse or mule around a fixed pole, much like a flower mill. The arastra was first used in the west by Mexican miners, but specimens have been found in other parts of the country and the world. When and where the first arastra was used is uncertain, but the device is ancient in origin.

The stamp mill was developed to speed up the process of creating small rocks out of big ones. The stamp mill came in all sizes and many of them still exist for observation purposes. Most mining museums will have an example for observation and there are many photos of them in use. Since they tend to be heavy and take up a lot of space only the serious collector will have one of these in their front yard. Still, a two or three stamp mill of small size makes an impressive bird perch for the front lawn.

Other types of ore crushers existed in daily usage, including small hand-operated compression jaw crushers similar to giant nut crackers and even smaller crushers for use in assay offices. Rotary crushers that used one grinding disk moving against a stationary cone or flat plate (much like an old coffee mill) were in use by small mines and assayers. These are highly collectible and can be found, although they are usually expensive.

Another type of crusher or mill is the rotary ball mill. These came in a variety of sizes and were an adaptation of a mill in use in other industries. These were very heavy and required a supply of high-grade steel balls. The ore to be crushed was introduced to the mill through a door on the side and the whole affair rotated at slow speed, allowing the cascading steel balls to break down the ore. The process is very thorough and can reduce ore to talcum powder fineness if required. The down side is that rotary ball mills are hard to maintain, very large, and very noisy. What can occasionally be found today are quantities of "ball bearings" that are in reality milling balls. These make great marbles for kids and super worry beads for adults.

Compressors and Classifiers

Sometime during the 1870s or 1880s, the steam and later air-powered drill came into being. The compressor and accompanying drills became the best way for a miner to bore holes in rock for the powder man to set his charge. Compressors were initially large ungainly items located outside of the mine. The compressed air was sent down to the miner through miles of pipe and hose, where he put it to use in the rock drill. They had the additional benefit of supplying fresh air to the miners at the working face. There are a surprisingly large number of these iron monsters laying around in the weeds in mining areas, many of which can be purchased for small change. Like their cousin the steam engine (actually a steam-powered motor as opposed to a locomotive), there are collectors who repair and rebuild them. Initially using belts and pulleys to transmit motion from the steam engines to the compressors, most were converted to some other source of power, such as an internal combustion engine or early electric motors.

Classifiers are pieces of equipment that often defy identification. In usage, a classifier is designed to sort large pieces from small pieces and usually to send the big pieces back to be reduced to smaller sizes. Many items in this class of iron look like Rube Goldberg creations and many were unsuccessful. Smaller classifiers are known that would have been used in the assaying process; these will show less wear and tear than major rock classifiers. One such large classifier still in use today is known as a "Grizzly." We honestly don't know why.

Drills and Bits

Drilling in mines is done to make holes for explosives to blast loose the rock. All early drilling was done by hand using progressively longer lengths of drill steel and arm power. Many examples of drill steel exist; while interesting, they are not expensive. As soon as the equipment to mechanize the drilling process using steam and later compressed air became available it was put to use replacing the strong arm of the miner.

The actual drills of one hundred years ago are little changed from the ones in use today and have little collector interest, except where they can be directly attributed to a famous manufacturer, such as Sutro or the Virginia and Truckee Railroad.

Molds and Smelting Equipment

Molds are one of the real survivors of the mining process. Because they are made of very hard and well-seasoned iron or steel they tend to survive the elements even when left out for decades. Once a mold has been made it really has little other use and consequently can typically be found in reasonably good condition. Many different sizes have been observed, from large "Dore" bar molds for casting lead/silver ingots weighing eighty-six pounds to small one ounce assay molds used only in an assayer's office. Finding a mold block with multiple sizes is an exception and these will bring premium prices. Variations exist in this area; one such is conical shaped molds used when a raw sample is being smelted to create a "button." These are simply known as "button" molds.

Other smelting equipment will occasionally be found, but typically items such as burners or furnaces are in bad shape and not worth much. The exception to this is the assay furnace, which can bring a good price when discovered. These are unusual because they were almost always used hard and their survival rate is low.

Ore Carts

In some areas, such as the mountain west, it is considered a real status symbol to have a mine ore cart mounted on your front lawn setting on a short length of rail on fake ties. Ore carts can be in almost any condition and still have some collectibility. In fact, ore carts were made to be rebuilt when worn out and virtually everything on the carts is replaceable, including the wheels. Finding a supply of correct parts is another story! Carts were made in a surprisingly large number of variations, styles, and uses. Different styles that are commonly found are end dump, side dump, tilting or rotating dumpers, carts made to be hauled by mules, carts made to be hauled by electric trains, and so on. They were made in all parts of the country and the best ones were made with steel frames and cast-iron hardened wheels. Early carts were also made using wood for the frames, sides, and even the wheels, but these did not last long in use and will seldom be found in one piece.

For purposes of illustration take a good look at the photos in *This Was Mining in the West*, which shows a wide variety of carts in use. Ore carts were commonly bought and sold and moved from one mine to another in a different part of the country. The largest and most common carts were the one ton size. These would hold a ton of rock or coal and were used in most mines at one time or another. A "mucker" or loader had to be able to load sixteen carts holding a ton each during a working shift before he could graduate to a less arduous position. There were other, smaller sizes and they will bring a little less on today's market, even though they are slightly less common.

To possess a whole train of carts would make a large and unusual lawn ornament, but some collectors do just that. Mining museums usually have at least one locomotive of one sort or another that was used to pull the carts.

Trains

Trains that have been used inside underground mines are unusual, because in the older mines the main motive power was by two- and four-legged creatures; some mines used mules, while most used men to push single carts to a point where the ore could be moved into a skip or bucket to be taken to the surface. Large ore buckets were also used and these will occasionally be found, usually in poor condition. The buckets and skips were both used to move men into the mines. A skip is a flat rectangular-shaped bucket designed to haul ore up an incline to a dump or tipple on the outside of the mine. They normally had wheels to ride on the rails of the incline and are fairly rare to find today. In the case of the Empire Mine in Grass Valley, California, special crew carts were used to move a large number of men down an incline shaft into the bowels of the mine. In the Comstock Lode mines the shafts were primarily vertical and they depended on specially designed lifts to move men into the mines and carts of ore out. The elevators were built to hold one or two carts on rails in the floors of the cages. The full carts were wheeled into the lifts, which then raised them to the surface where they were moved to a tipple or dump, depending on the contents of the cart.

Trains of larger ore carts were moved from the mines to the mills, sometimes miles away. The mines used steam locomotives from the earliest and made many ingenious small engines to accomplish this task. Because they were steam powered, these engines could not be used inside the mines and as soon as electric motors were developed that were powerful enough to move cart trains they were incorporated into low-slung electric engines that could also move carts outside.

Pelton Wheels

The Pelton wheel, so named because of the person who patented the invention, is merely the refinement of a very old process for creating rotary motion from moving water and gravity. In California and elsewhere, Pelton wheel engines were made in every size imaginable, from about four inches in diameter to monsters thirty feet in size. The largest were capable of producing up to an estimated 5,000 horsepower. Pelton wheels depended on a reliable supply of water and some way to contain it to direct the water into the cup-shaped cogs on the wheel. Enclosed wheels were used inside mines to produce electricity from falling water in pipes that had been pumped to a higher level by large steam-powered Cornish pumps on the surface. Such was the system used in the Comstock Lode mines to produce electricity for lights and other tools at the lower levels.

Outside the mine, very large Pelton wheels were used in place of steam engines to power pumps and generate electricity.

They were largely used to produce electricity, but were also harnessed to air compressors and hoists. Today, Pelton wheels of older vintages are much sought after and small finely made ones will bring rather premium prices. Large wheels are best left to museums to display.

Retorts

Retorts are items not commonly found, for as much as any other reason because people frequently do not know what they are. These globe- or pineapple-shaped iron bombs are in fact what was used to cook off the mercury from a gold recovery process called amalgamation. In practice the mercury—much of which was mined in California and Nevada by the way—was placed in the sluice boxes or the recovery system of a mill. In stamp mills copper sheets were coated with mercury that then chemically "grabbed" the gold as it passed over it. The result was a gray material that was called "sponge." This sponge was placed in the retorts, which were heated to a high temperature. Since mercury reverts to a gaseous state easily, the mercury gas was driven off through the pipe, leaving the retort, and recovered in a bucket of water, where it was easily handled with a spoon or cup. What was left in the retort was impure gold. Mercury was bought and sold like any other commodity in iron flasks, which occasionally turn up and have some value unto themselves.

The iron retorts can be found whole or in halves, but only a complete retort with the cooling pipe is really worth having. Artifact retorts that have been blown up are occasionally found because they had an annoying tendency to explode if overheated. Caution needs to be used when handling the retorts, but as long as gloves are used contact with any remaining mercury is unlikely.

Sluice Boxes

The sluice box of Hollywood movie fame was most generally made of wood and was in fact a very inefficient way to recover gold. The actual boxes used were made in whatever size was convenient for the task at hand. The longer they were the better they were at recovering gold from the stream beds. The sluice box of the gold rush was not capable of recovering much more than nuggets, since they leaked water badly and did not catch much of the very fine gold. To find a real period sluice box today would be pure luck and authenticity would always be in doubt. Later on metal boxes were developed and while more efficient, they were quickly put aside in favor of mercury-based amalgamation processes.

Hydraulic Monitors

Monitors were huge water nozzles used to direct enormous quantities of water in a concentrated stream against a hillside to move mountains of overburden and get to the gold-carrying pay dirt near bedrock. The monitor or "Little Giant" was a nozzle that was the end of the line for water that had been collected in the mountains and was conveyed through a complex system of ditches and conduits to the mine diggings. The water was then forced into smaller and smaller pipes until the end was reached. The end was the monitor, which put out a steady stream of water some three to eight inches in diameter with enormous energy. When directed at a hillside made up of dirt it quickly broke down the mountain into a highly viscous mud. The slurry at first contained little gold, but as the water jet was worked lower and closer to bedrock more and more gold was moved along with the water and mud.

Eventually the dirt hill was completely destroyed and while the entire operation was funneled through a long system of sluice boxes huge quantities of gold were recovered. The down side was that when entire hills were washed away all the silt had to go somewhere, and the somewhere was into the region's waterways, which quickly became choked with muck. By 1884, hydraulic mining was legally outlawed in California, even though the process was used effectively in other areas of the US into the thirties or forties of the twentieth century.

The huge nozzles can be found occasionally and they will normally have an associated weight box and base. To be complete all the parts need to be together, although nozzles can be found, as well as the tapered funnel and connecting angled pipes. Pricing of these items is a problem.

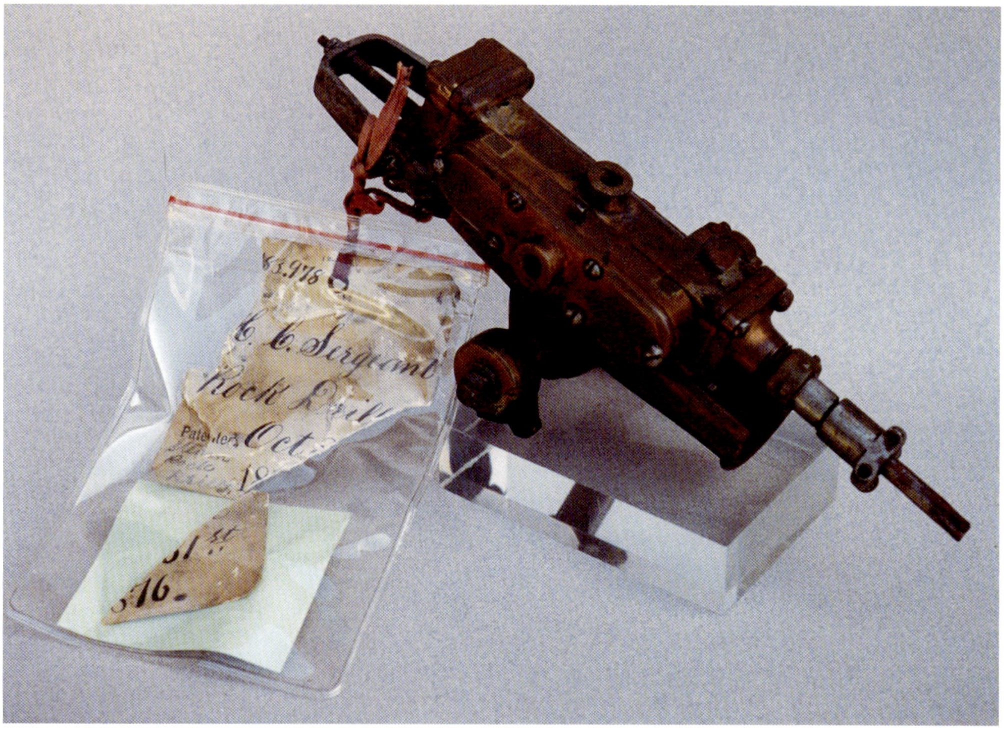

An actual patent model of a rock drill. Patent was issued to H.C. Sargeant No. 183,978 dated Oct 31, 1876. AM

This large button mold is cast iron and was made in Los Angeles by the Braun Company. This mold stands about eighteen inches high, is quite heavy, and was used to make a first pour of molten unrefined ore. The mold would be turned over and tapped to release the cone-shaped metal button. The base or point of the cone could then be easily broken off from the slag and would be almost pure metal. $125 to $250. RB

This tag is all that remains of a hydraulic "Giant" as made by the Hendy Company. Dating from pre-1890s. The tag alone, $100 to $400. SB

Two Pelton wheels produced by the company of that name, both in good condition. The larger of the two is about twenty-four inches in diameter. Prices vary widely, $500 to $2,500 each. RB

A small Pelton wheel built in San Francisco. About eighteen inches in height, it is in fine condition. $600 to $900.

A display of Pelton wheels of different sizes. Located in Placerville, California, this display ranges in size from eighteen inches to one of about ten feet.

A small production or sampling rock crusher. This example could be turned by hand or driven with a belt from a separate power source. A very handy item to own if you have a hard rock gold mine. $450 to $600. RB

A grouping of smelting ladles, all different and all well used. $75 to $175. RB

This rusty, plain looking cylinder is actually a mercury flask and little different than those in use today. Mercury is very heavy and this flask would hold about 275 pounds. $30 to $50. RB

Two large ingot molds of production size. They could be used for either gold or silver. If marked, $300 to $500. RB

This plain looking wooden bucket wrapped with iron straps is actually quite rare. This would have been used in a small mine to bring ore to the surface by a windlass to be smelted. In good condition, $300 to $650. RB

This innocuous looking metal box is a "high grade" box used in a gold mine to collect very high grade ore for movement to the surface. $175 to $250. RB

This rather strange looking device is a hand-cranked single stamp mill for use in assaying or by a small miner. In unrestored condition, needing only clean-up and replacement of the handle. Approximately $750. EC

A mine "skip" cart that would have been used where the entrance shaft of a mine was on an incline. In use the cart would be filled with ore and run to the surface on rails. Once at the top of the workings the cart pivoted into a chute opening, dumping the contents into a tipple for crushing. Actually quite rare, this cart held about a ton of rock. In good condition, $800 to $1,000. EC

A good selection of ingot molds. The largest is for silver or gold, while others at the top are assay molds to test batches. $150 to $300. EC

This four-stamp mill was restored by its owner and sold for $2,000 to $6,000. EC

What looks like a pile of junk are actually parts for mercury retorts, including one held up in the left center of the photo. Values for complete units, $125 to $300. EC

This painted pink mine cart was made by the Joshua Hendy Iron Works of San Francisco, California (see detail). $2,000. The ore "skip" is from the Empire Mine in Grass Valley, California. About $1,000. EC

The scruffy looking white ore cart, $950. Round bottom "Mule dump" cart, $1,300. EC

A pile of production molds, mostly lead/silver dore bar molds. The rectangular molds in the front would have been used for casting silver/gold ingots for movement to the mint or to a smelter. $200 to $1,000, depending on markings and condition. RB

The bucket in front is a tram bucket, complete with the carriage on top and the tipping handle on the bottom. Smaller than many tram buckets, a whole line could move a good volume of high grade ore a significant distance in a short time. About $175 to $300. The round bucket in the back is a common shaft ore bucket, $300 to $500. RB

Most early machinery manufacturers produced plates of bronze or brass to identify their work. $65 to approximately $400, depending on age and condition. EC

Another Truax ore car, this one made by Globe Iron works in Stockton, California, and sold by Nevada Hardware in Reno, Nevada. Ore carts were treated roughly and those with the manufacturer tags still intact will bring more when sold. $1,000 to $1,600. RB

A large foundry ladle for moving molten iron from the hearth to a casting location, also called a slag cart. Not normally used for gold or silver, this makes a perfect flower planter. $400 to $600 if complete. RB

The wheel barrow to the left is all steel and was used inside a mine to move ore. Approximately $150. The large casting on the right is the base of a small stamp mill and was cast by the Virginia and Truckee Railroad foundry in Carson City. Marked. $2,000 to $3,000. RB

One-ton ore cart, complete. $1,600. RB

Mine cage on elevator. $2,500-$5,000. RS

Joshua Hendy Builders plate, San Francisco, California. $500. DK

A 3-inch Pelton Wheel gauge. $150-$300. BS

Brass Pelton Wheel plaque, 8 x 17 inches. $250-$300. BS

Two ore cart bodies, usable if you had trucks to put underneath them. $250 to $400 RB

A brass tag on a Truax Silver State ore car made by Truax in San Francisco, California.
Tag alone, $400. RB

Photo taken by the author in 1998 of the remains of a large compressor near the mining camp of Cambridge, Nevada. The mine shaft has been imploded just to one side of this large chunk of iron waiting for someone to haul it off. The age of this iron monster is unknown, but it had a very large engine mounted on a truck-type frame with solid rubber tires. Value is the price of bail. DP

A tram bucket that would have been suspended by a cradle on a long rope or cable. $350 to $500. RB

A pile of mercury flasks. These iron flasks have only minimal value to the collector. $50. RB

A good-condition maker's brass label from an ore cart made in Stockton, California. The tag, $150 to $400. RB

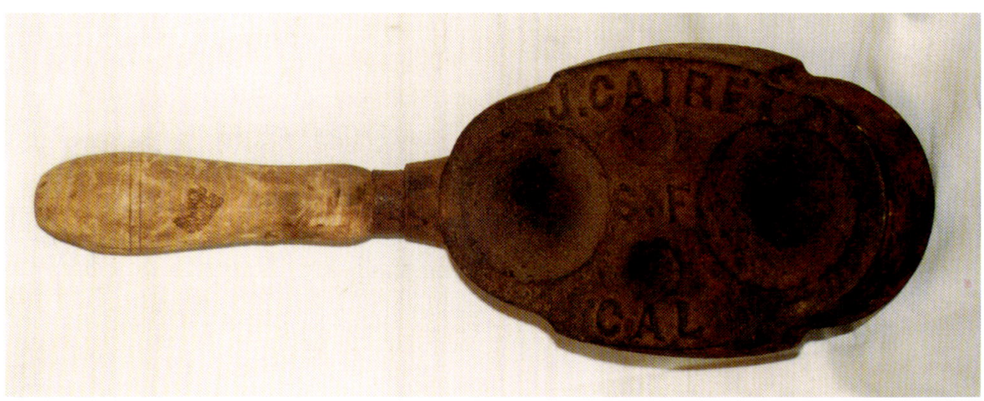

A good-quality button furnace mold from the J. Caire Co. of San Francisco. $140 to $180. RB

Drill steel comes in different lengths. This one would be a "starter" piece and is almost new. $35 to $50. RB

The business end of a monitor hydraulic giant that was used to wash overburden off pay dirt. This one is missing parts and is actually mounted on the front lawn of the Sutter Creek, California, town hall. $1,200 to $3,000

Another "Monitor" hydraulic giant of a larger size, missing the counterbalance weight box and the director's arm. This is located in a museum in the California mother lode district. Approximately $2,500

A diesel-powered shay or geared locomotive. These were common around mines in both steam and diesel power. Seldom found except at museums, they are heavy and tend to take up a lot of room.

These two complete stamp mills are set up at a museum in Placerville, California. The one on the left is the more common type with five stamps. The one on the right is unusual with only two stamps. Both would have been powered by steam engines. $2,500 to $10,000.

Ore cart found in a backyard. This one has small sections of steel track to set on. The track sells by the foot and depends on the length for value. $1,500. RB

Chapter Thirteen

JEWELRY

We decided to include jewelry as part of mining collectibles because of the large number of items that were made and sold during mining's heyday.

RINGS are a less obvious item of jewelry, but some rings are directly attributable to the mining boom. Items such as nugget rings are obvious, but the gemstone cut known as an "Old Miners" cut is less well known. In a few areas of gold mining "jewelry gold" was common and the early miners high-graded such nuggets and ore samples with astonishing thievery. These display nuggets were sold at inflated prices and today occasionally show up as objects d'art. Rings were frequently made using smaller nuggets (indeed these rings are still made and sold in Alaska) and these will bring premium prices for better-quality items.

FOBS and CHAINS were common in a time period when a man of substance would have a gold pocket watch with a fob and chain made up of nuggets. Typically such items were passed down from father to son, and to find one today is unusual. Prices for these items are high. Be cautious when buying one of these items, as they are still being made and sold to followers of old western Americana.

STICKPINS were jewelry items that were worn by both women and men, who used them to adorn a tie or as a lapel decoration. Stickpins were made using gold nuggets and gems or a combination of both. Since they have fallen out of fashion the collector is the only person who uses or buys them anymore.

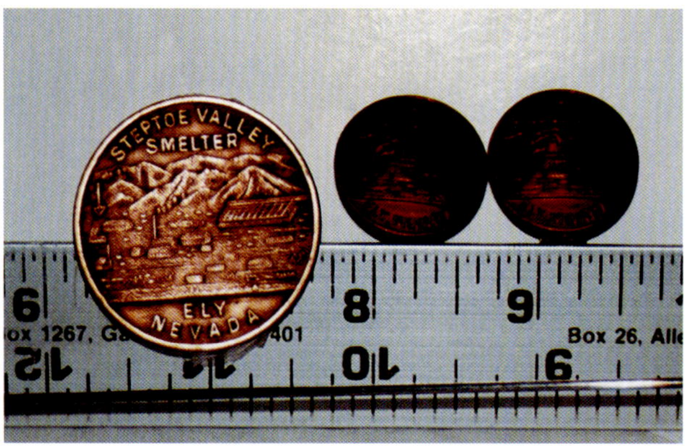

Earrings, a brooch, and a hatpin, all with the Steptoe Valley Smelter logo from Ely, Nevada. All items are in perfect condition and have the same picture. Ely was a large copper producing area from the Ruth open pit mine for many decades. $45 to $100 for all three. RB

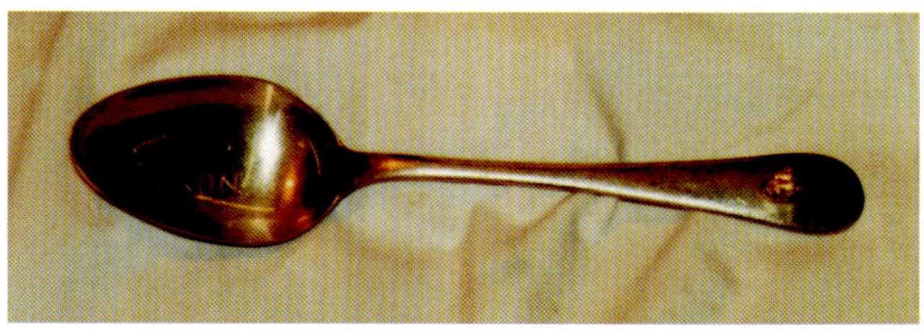

Souvenir spoons were just as popular in the mining camps then as now. This nicely crafted spoon had thin gold plating over silver and is engraved "Yukon" with a gold nugget. $50 to $200. RB

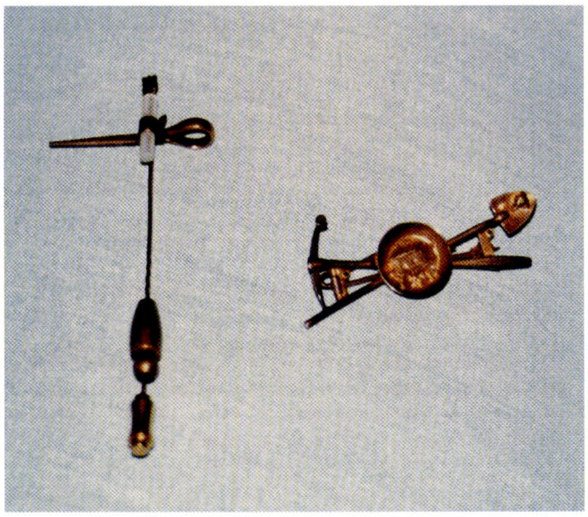

These two period pieces are women's pins made in the form of miners' tools. The miniature candlestick holder has a candle made of a hard white substance with a flame of red stone and is a Colorado School of Mines graduation pin. $500 to $750. The other pin is crossed tools with a nugget in the gold pan. $150 to $200. RP

Rough gold nuggets were frequently welded to copper wire to make stick pins that were worn to show off a person's material wealth. The long item in front is a hat pin with a nugget of what is believed to be silver ore. $200 each. RB

This medal was apparently an award for some sort of life saving or medical emergency. It is dated 1926 on its reverse and was given out by the Department of Commerce Bureau of mines, San Francisco, California. It would appear to be gold plated brass or copper. RB

Chapter Fourteen

KNIVES AND GUNS

Guns were as much a part of everyday life in some mining camps as lunch buckets. It was fairly common for some early miners to carry a pistol or have a shotgun close by to keep bandits away or to ward off the occasional bear or skunk. But it is part of western lore—largely because of Hollywood—that every miner had a Colt on his hip and a Winchester close by. Nothing could be further from the truth. The most common collectible in this area is the double barrel shotgun. The earliest shotguns were muzzle-loaders, but as cartridge-firing arms became available the front-stuffers were quickly relegated to the closet. Ownership of rifles and shotguns was common and they were in use on a daily basis to supply food to the early miners and to the emigrant trains that headed west.

In other parts of the country different kinds of guns were common among certain societies, but the average miner who went underground at dawn and returned after dark was hard pressed to keep beans on the table, much less buy a pistol that could cost a month's wages. To be sure, many men (and women) bought pistols or rifles when they became more available and cheaper. It should be remembered that at the beginning of the western gold rush, the classic Colt Peacemaker was some twenty-three years in the future.

But there were many cheaper guns kept and these are highly collectible today. Items such as the early single-shot percussion pistols, pepperbox pistols, and percussion muzzle-loading shotguns are all highly collectible. There is even a very active market for "dig-ups"—artifact firearms that are completely unserviceable.

The top of the heap of firearms collecting of this period is to find guns that were carried by Pony Express riders or used and marked by the Wells Fargo Company. Mines frequently bought and kept pistols and shotguns for guarding gold shipments, and if found and authenticated these bring premium prices.

Knives were very much a part of everyday life and knives both large and small were in common use in all mining areas of the country. Miners frequently carried large "Bowie" knives both for eating and for prying nuggets out of ore. A good knife was one of the more utilitarian tools in a miner's day. There are a number of good books available to identify and value old knives so we will mention them only in passing here.

A very rare factory short-barrel Colt, model 1861 Navy with factory engraving and ivory grips. This Colt came out of Bodie, California, many years ago. $15,000. RB

A percussion pistol by Wesson that attempted to circumvent Colt's patent. Actually a very good mining camp collectible, the value is considerably less than the Colt. $500 to $700. BM

A very nice-condition Sharps derringer with four barrels in .22 RF. These were very popular with gold rush miners and gamblers, as well as ladies. The stag handle is not original, but the pistol is otherwise in very good condition. Shown with a nickel-sized gold in quartz nugget for scale. $500 to $600 RP

Knives were carried by almost everybody in the gold fields and used for everything from eating to prying nuggets out of rock. This is a common brand called the "Argonaut," along with a period sheath. Most of these knives were British-made specifically for the California trade. $300 to $400 in unaltered condition. RP

Authentic Allen's Pat. pepperbox of about .36 caliber. $1,500 or more. RB

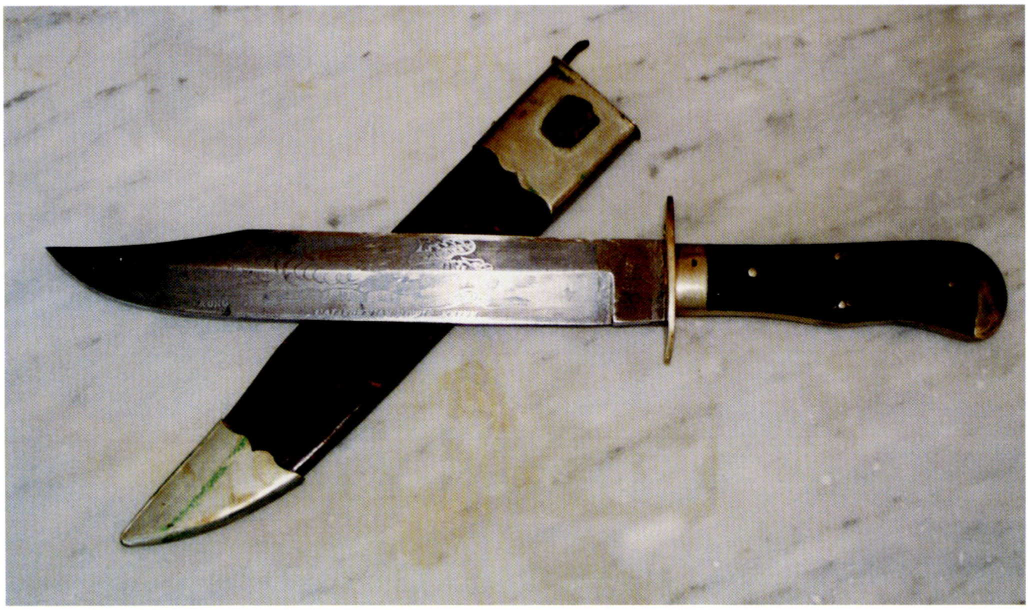

A large period gold rush era bowie knife with an etched blade reading "California Bowie Knife" with an eagle and other figures. This type of knife was extremely popular in the early days of the western mining boom and is shown with the original scabbard in excellent condition. About $5,000. RB

A fine-condition gold rush era knife with the inscription "Lou Howard" on the blade. Handles appear to be of bone or ivory construction. Similar knives have been sold for $10,000. RB

Chapter Fifteen

LUNCH PAILS

The ubiquitous lunch pail came in all sizes and shapes, but all were designed to do one thing: hold a hearty meal for the hard-working miner who had to eat where he worked. The pails were usually compartmentalized to hold a sandwich or a pasty and a tin of tea or water. Some had two liquid holders for tea and soup in addition to the solid portion of a meal. Miners worked hard and needed a lot of calories to maintain their bodies. Some of the lunch pails were meant to have an alcohol lamp placed under the bottom to heat the contents.

Many of the pails were made of enameled tin, while a type of "granite wear" is one of the more common types. Sizes run from very large to sandwich size. Miniatures of all types have been observed and these were normally not for sale to children as toys, but were salesmen's samples. Miniatures will bring prices at least equal to the full-sized article. The lunch pail is so common in some areas and the survival rate so high that in spite of their age some of these items price relatively low.

The lunch pail is a good item for new collectors of mining to begin with. It is available, not too expensive, and creates a good display. Size works against the lunch pail if you get into them to the extent of one photo shown in this book.

Miners' lunch pails come in lots of different shapes and sizes. This one has a compartment or tray in the bottom, as well as a top than lifts out and probably contained tea for the miner to drink with his meal. Prices vary with condition and this one is missing the small cap on the pour spout. $150 to $250. RB

These containers were used to provide the miners with a drink of water or tea while on the job. The tinned flask on the left has a cup on the top. $100. The copper flask on the right is more unusual. $200. EC

Lunch pails were common and well used. The left pail is a Western marked "Leadville Colorado." $4,000. The right one is unmarked, $150. EC

This black, porcelain-finish lunch pail comes with its own cup. $250. EC

The Royal lunch pail of granite wear type, $200 to $300.
The two small pails are salesmen samples and very scarce. $100 each. EC

Three-compartment miner's lunch pail of tinned iron. $75 to $200, depending on condition. RB

Miner's lunch pail with a top section that held liquid, usually tea. $125 to $200. RB

Miner's lunch pail where the top is a screw-off tin cup. $125 to $200. RB

Chapter sixteen

MAPS

Maps in virtually any condition are valuable to the collector because they frequently have information detailing the locations of long-forgotten mines or town sites. When these older maps are found one of the first things noticed is the high quality of the reproduction. Most maps of the 1870s and later were done using processes that are no longer available. Having said that, in many cases the maps drawn of some areas will sometimes be found to be inaccurate, so do not go to court with one. Actually, the ability of surveyors of the pre-1900 mining eras to accurately draw maps is a real tribute to their abilities.

In some cases, such as the Comstock Lode in Virginia City, Nevada, the only accurate record of where mines were is directly attributable to the mining maps done at the time. Since many of the mines are closed off from the surface now, period maps are the only way to know if a property owner is building a new hotel over a very large hole in the ground. Basically, the larger the map and the better the condition the more valuable it is.

Maps of especially famous boom towns will usually bring the highest prices simply because they are very rare.

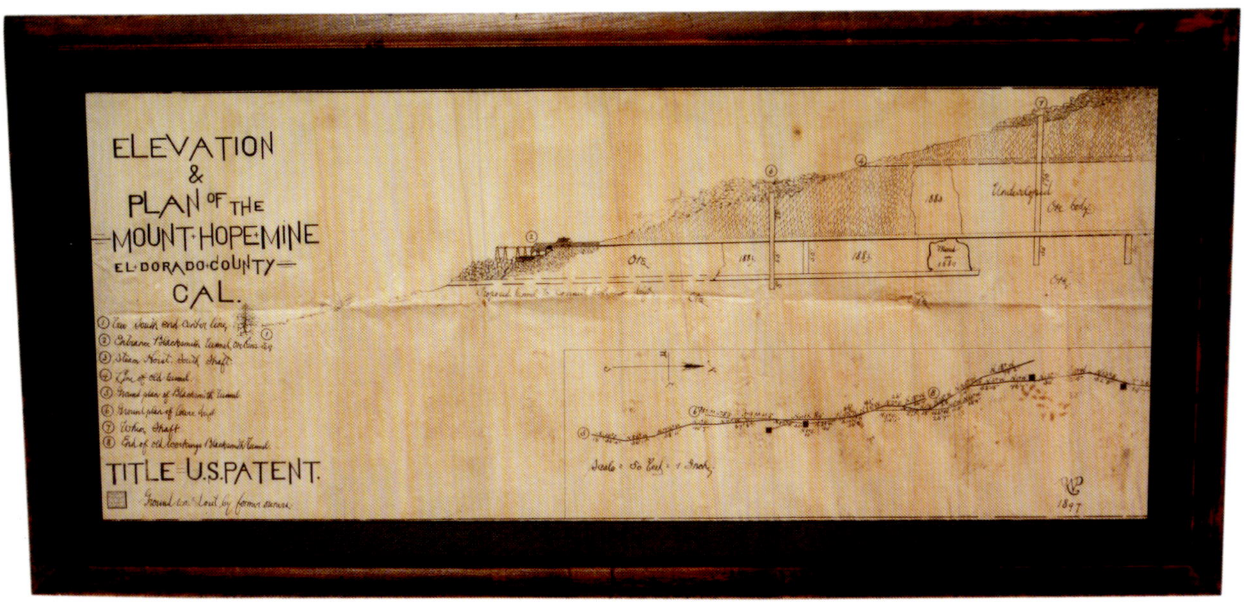

Maps of mines and mining areas were very important to the mine owners and the mining engineers. They provided a legal basis for claims and allowed geologists to follow ore bodies. This particular map is of a mine in the mother lode district of California. It is mounted and framed to protect it and shows evidence of folding and storage. In average condition, $500 to $600. HD

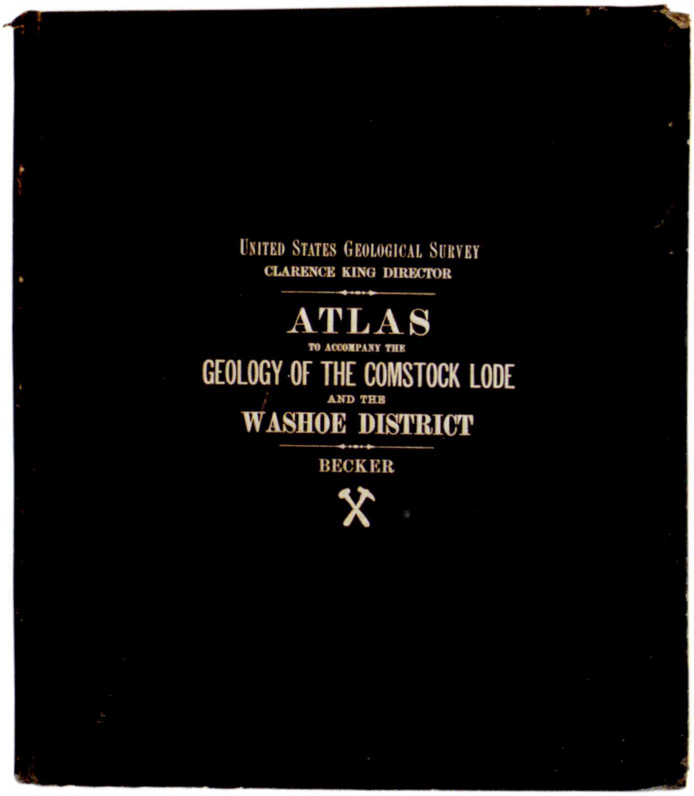

Above and right:
Produced in the 1880s, this USGS production is a book of maps and is extremely detailed. It was extremely accurate for the time and copies today are rare in any condition. RB

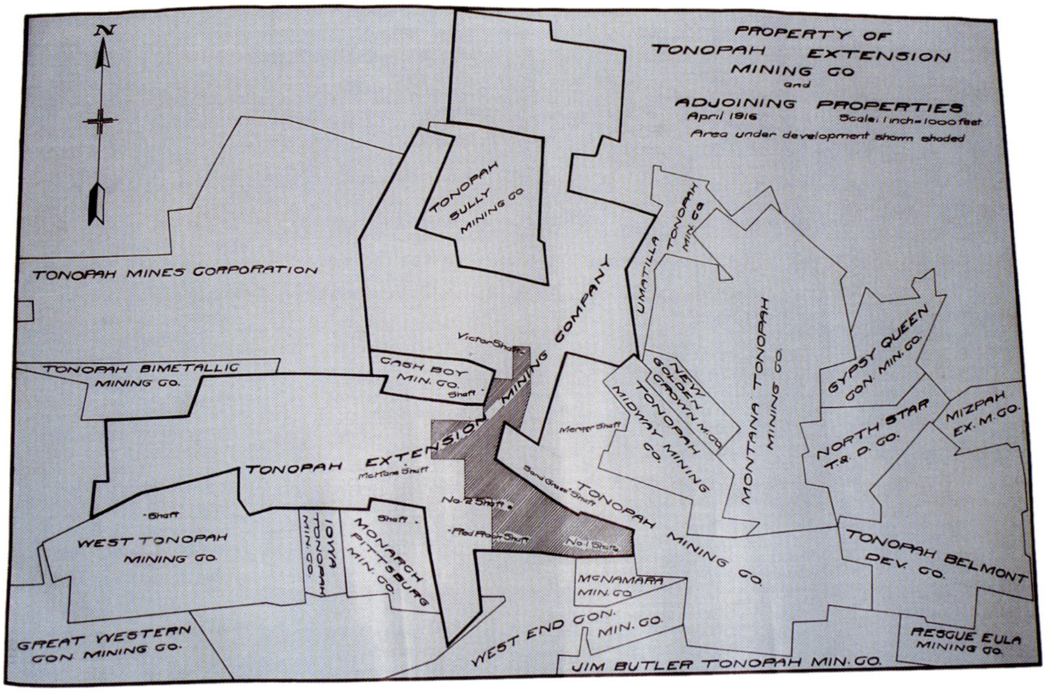

This finely printed map is unusual, as it was printed on waxed linen fabric.
From the Tonopah Nevada area, dated April 1916. RB

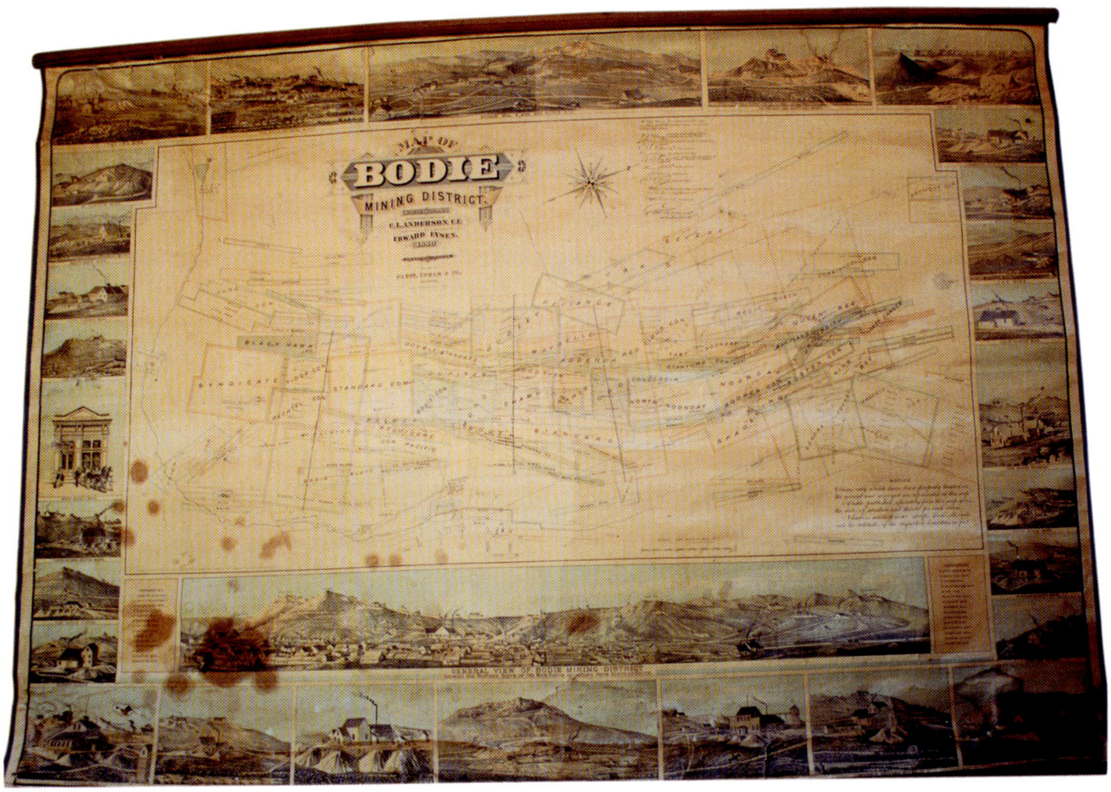

Maps from well-known or famous mining towns are the most valuable and this one is about as rare as they come. The ghost town of Bodie, California, was one of the wildest and richest of all western mining towns. This map, perhaps unique, is printed on oil cloth or canvas. It is about four feet by five feet and mounted so it could be hung on a wall. In tender but fairly good condition it is conservatively valued at over $15,000. RB

Chapter Seventeen

METAL SIGNS

Metal signs are one of the lesser areas of mining collectibles, primarily because they are really hard to find in good condition. The earliest signs used in mines were made of wood and probably have not survived the ages, except for those in the desert southwest. Metal signs were used virtually everywhere in the mines because they resisted the ravages of the mine's environment of dust, water, and abuse. The early signs were porcelain coated and of very high quality. Finding a sign that would have been used inside a mine of this age is difficult for today's collector. Later signs were enameled and nearly as collectible. Still later the signs were simply painted on tin sheets. Some of these will be found in all sizes and the more complex and larger the sign the higher the price.

This solid brass sign from the Sutro Tunnel Co actually has another sign painted on the reverse. It is quite rare. $1,000 to $1,200. RB

Fine-condition signs from inside mines are rare. $150 to $200 RB

Another inside mine sign with slight damage. $275 to $400. RB

Porcelain or enameled signs in pristine condition are tough to come by. $150 to $200. RB

This porcelain on steel sign is well used. $200 to $450. RB

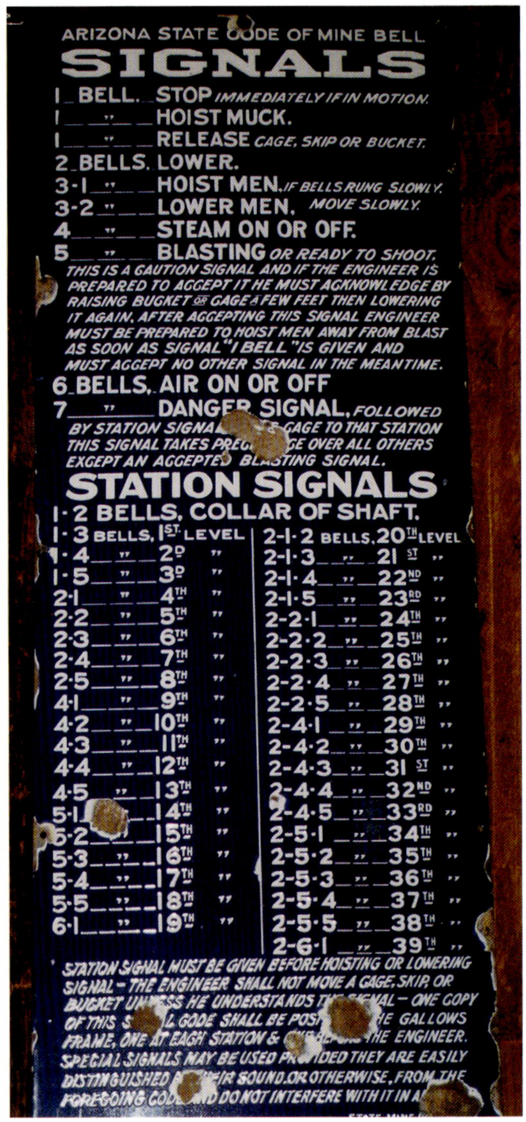

Full-sized bell signs came in different forms and this one in enameled steel is about five feet high. From Arizona, about $800. RP

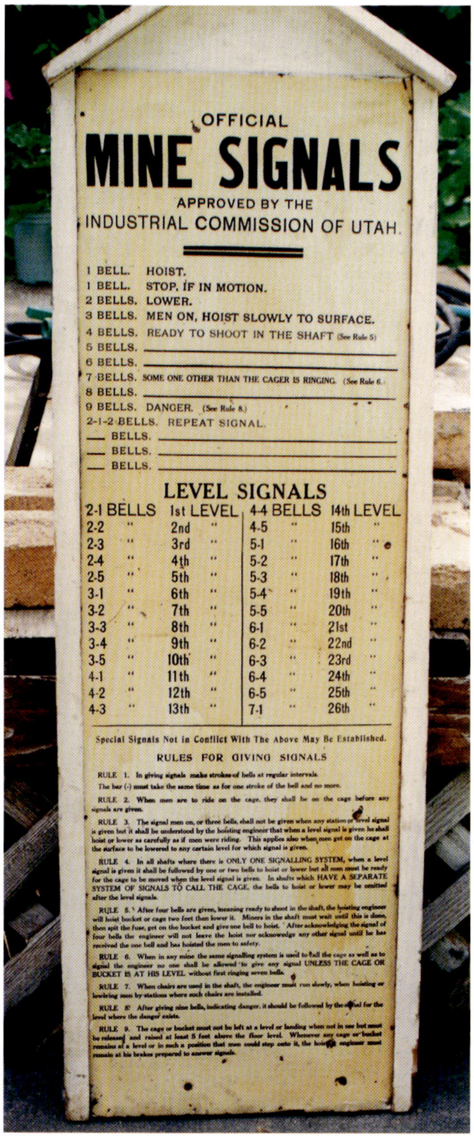

From Park City, Utah, this sign is mounted. $350. HD

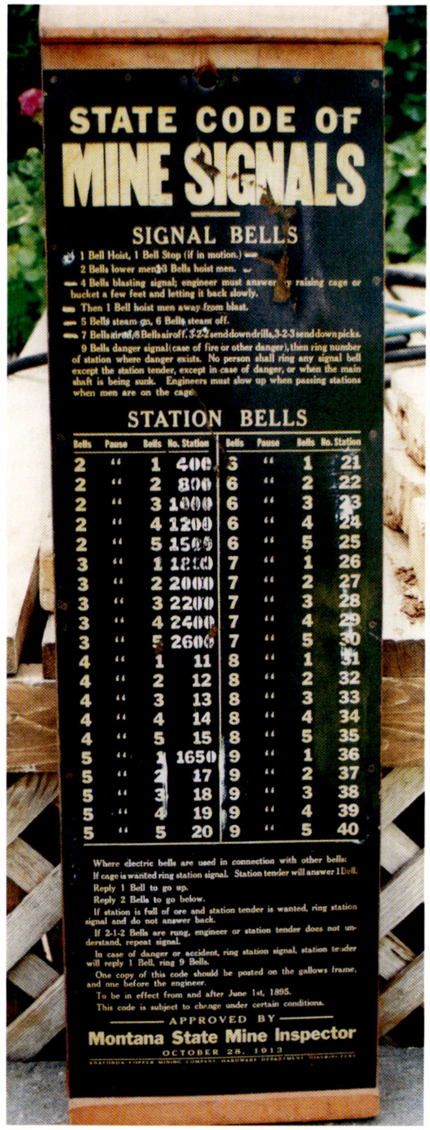

Another typical metal bell sign; this one is from Montana. $250. HD

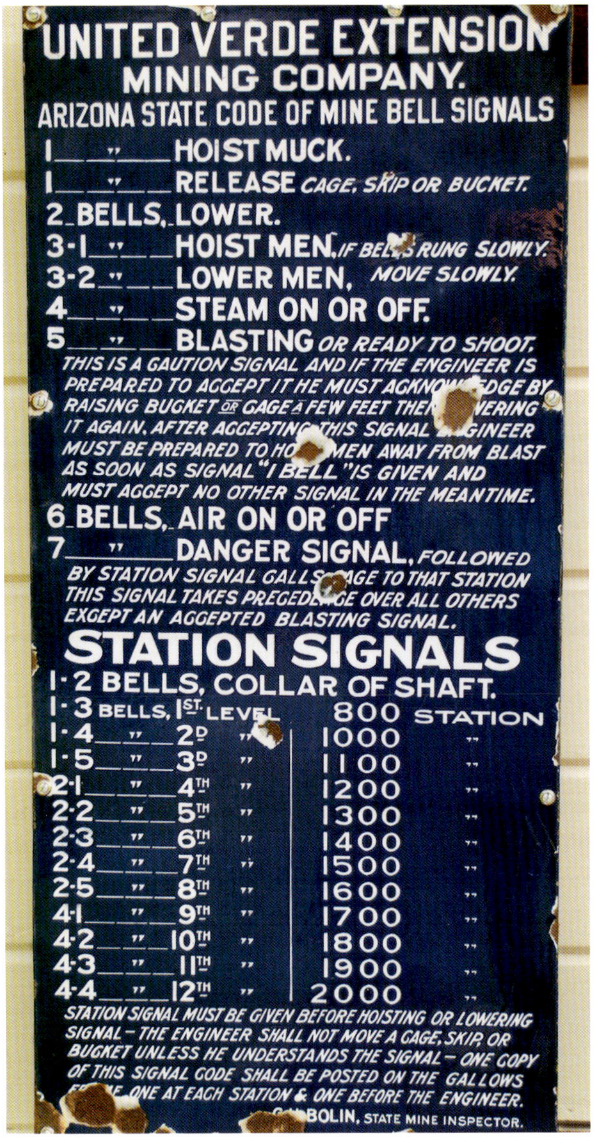

Another Arizona mine bell sign; this one is a bit worse for wear. $650 because of its rarity. HD

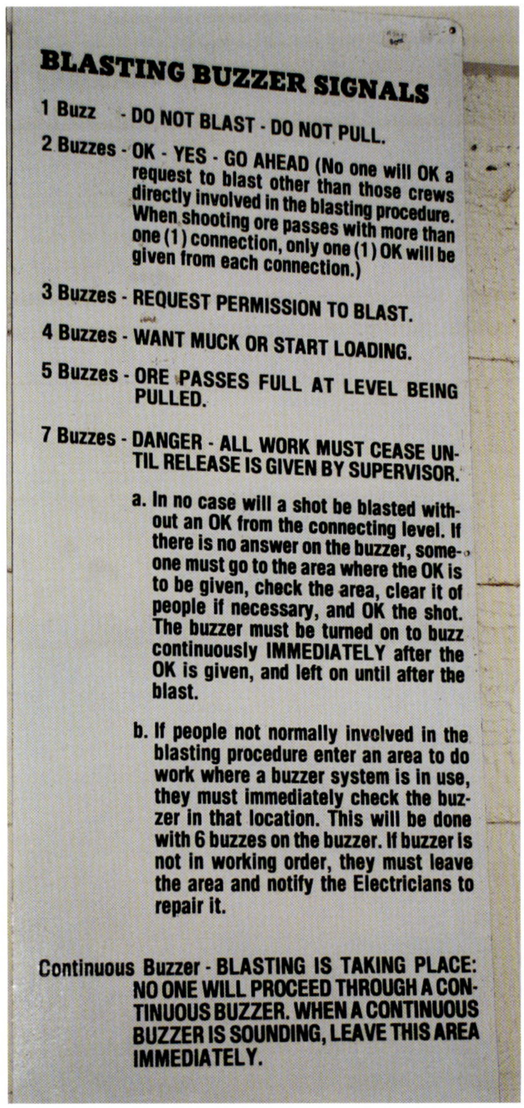

This nice white metal sign is not marked as to origin and therefore its value is less. $200 to $250. HD

Chapter Eighteen

OIL LAMPS

Oil lamps were the successor to the candle and the precursor to the carbide lamp. The lamps took a variety of shapes—some quite small—and transitioned from the stick holder style (very much like a miner's candlestick holder) to wear on the front of a hat. Oil lamps used a wick inserted into the oil held in the base of the lamp and simply burned to provide light. In use they were not a real improvement on candles and were dirty, sooty, and bad smelling, since almost any type of animal fat could be used. The light they provided was generally not good and the miners were forced to carry small flasks of oil with them to refill the lamp when it burned down.

Lamps of all sizes are highly collectible and they may be found in both brass and steel or tin construction. Brass is more common than steel and new lamps in the box can bring premium prices. They appear to have been more common in western mines than in the eastern coal fields, but evidence exists that they were well used there also. Rarely, lamps with two wick spouts will be found and these will bring premium prices. A good example of their usage can be seen in the Sean Connery movie *The Molly Maguires*, which was set in the eastern coal fields.

There have been at least two hundred one different oil lamps made through the years (probably more) from an unknown number of makers. Some lamps were sold by retailers who did not in fact manufacture the lamp, making identification all the more difficult by the collector.

This oil lamp came with a carrier and a candle and was a patented item. Value is high, $500. EC

These oil lamps are a cross-section of different types and manufacturers. From left, clockwise: $200, $75, $150, $75, $150, $1,000 (has an integral spike for sticking in a timber), and $35. EC

Oil lamps made of brass are generally higher in value while tin lamps are more common and lesser in value. From left: $75, $75, $200, $150, and $75. EC

An interesting group, this consists of the full size article with miniatures of both types. The miner's safety lamp on the left is $250. Its miniature, $120. The oil lamp, $125. The miniature, $250 to $300. EC

There is a lot of material in this photo. Prices are for full boxes:
Alco Burners (wicks) $30.
Baldwin Lamp $50.
Justrite boxes $30 each.
Anton Lamp Wicks $275.
Hardsocg's miner's pants $500.
Justrite tips $150.
Powell's Safety Squibs $85. EC

Miners' oil lamps in boxes are quite rare so prices go up accordingly. $150 each. EC

Oil lamps also came in different sizes. Two larger lamps, $35 to $75 each. EC

Oil lamps came in many different shapes. Front left: copper and half round, maker unknown, $300; Williams, $400; and Williams, $500. The lamp behind with the sticker holder, $500; without the holder, $300. Husson, $1,000. EC

The double-outlet oil lamp, $150. Right, $50. EC

Oil lamp holder for use on a hat or as a sticker in a timber. Made by the Husson Co. The sticker by itself, $500. The lamp, $250. RB

An illustration of one collector's oil lamp collection. The large lamps on the left are sometimes called "mule lamps," as they were hung on a mule's harness for illumination. The large flange was intended to keep heat and flame away from the animal. HD

Referred to as a mule lamp, this oil lamp stands approximately eight inches high. Note the large size of the hook on the back, intended for putting on an animal's harness. HD

Every collector dreams of finding one of these. This is an unopened box of Hardsocg miners' oil lamps. Complete, $1,000 or more. HD

Chapter Nineteen

ORE SAMPLES AND NUGGETS

In some respects, collecting ore samples is a laborious process, but the collecting of nuggets is an exciting hobby and one that can be quite profitable. Ore samples can be difficult to come by unless you happen to own a rich gold mine. In fact, very few mines actually ever produced jewelry gold nuggets or ore samples worth putting on display. In times past large nuggets were found mainly in placer deposits, where they were tumbled by water action and were consequently rounded in shape. The rougher the nugget the closer it would have been to its origin. Underground hard rock mines seldom produced or produce collectible gold nuggets, except as found in veins of quartz-based rocks. That we know of today, the Goldfield, Nevada, mines were early producers of "jewelry box" gold samples that were high-graded or stolen and sold to the public, rather than being melted down for bullion. During present times, the famous Sixteen to One Mine in the northern California Mother Lode district still pulls out large examples of crystalline gold and these sell for premium prices.

Prime examples of ore should be identified as to origin and include a label detailing the level where the sample was collected. Dating of the sample is important, as serious collectors show little interest in samples after the 1930s.

Valuations for collectible nuggets and ore samples are largely dictated by what a seller asks and a buyer is willing to pay. The bullion value of the gold has some effect on the price paid, but is not much more than a starting place for determining the real worth of a gold nugget or ore sample. In essence, anything over one-quarter of an ounce of gold per nugget will bring a 50% premium over market value. As the nugget size grows so does the premium paid.

Actual ore samples tend to bring less than nugget prices and will depend on much subjective evaluation. We should also caution the buyers of nuggets to beware of fake nuggets, most commonly made from old school rings. Real gold nuggets have a certain look to them that is different than a fake one and only with experience looking at the real thing is it possible to see the difference. If in doubt ask questions. A good item to have handy is a 24 Karat gold coin or ingot to check the color of a nugget. Most nuggets have a ninety to ninety-five percent gold content and will be just slightly off color when compared to a coin.

There are chemical tests that can be run on nuggets to determine whether they are fakes that are not destructive to the nugget. Something called a touch stone is utilized that takes a small scrape off the nugget. The scrape is then tested with acids and the reaction is noted. The experienced chemist or assayer can quickly tell the purity of a gold nugget. Gold nuggets that are mounted on rings are frequently plated with 24 K gold to enhance their look and this process will destroy the collectible value of the nugget. Beware of rings with nuggets that appear to be cast into the ring. This usually means the nugget is in fact a casting and hence a fake.

Ore samples that contain quartz or other rock can be mounted on a wooden platform or mounted in a box. The value is not affected by either display method. Samples that contain other types of metals, such as silver or copper, are more difficult to keep shiny because the metals tend to quickly oxidize and tarnish. When this happens the attractiveness of the display is diminished. Cleaning with commercial agents is discouraged and nuggets may be found with a coating of lacquer or clear varnish to preserve the natural luster.

Gold in quartz nugget. Value is dependent on the current market, as it is too large for jewelry. The five cent coin is shown for size comparison. $300 to $500. RP

A cut slab of gold ore from the Grant Mine near Fairbanks, Alaska, a hard-rock producer of gold for many years. This sample has a high gold content and is labeled where it was taken from inside the mine. $200 plus. The Colt SAA above is shown for size comparison and is a 1970s production, $1,000 in today's market. DP

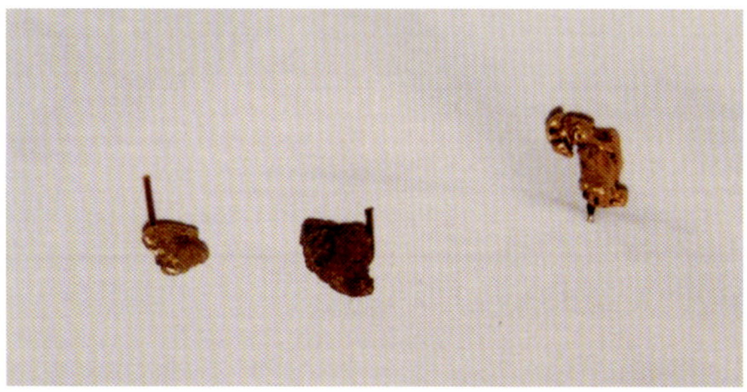

Three gold nugget stickpins shown about actual size. Values are hard to determine because these items are out of fashion. About $250 each. RB

This is a large chunk of Galena ore that is high in silver. The color is actually a result of flash reflection. This is very high-grade material. HD

CHAPTER TWENTY

PAPER

Paper products are a large field of collecting that covers many items directly associated with mining. We will discuss and illustrate catalogs, checks, deeds, government publications, maps, script, stocks, and perhaps more.

Catalogs

Much of what we know today about mining equipment comes from surviving catalogs produced by manufacturers and suppliers to the trade. It should be remembered that one hundred plus years ago, no electronic media existed and advertising and sales of equipment was done by personal contact with traveling salesmen using catalogs of what they sold. These catalogs were all important to the mining camp operators, who would often order items sight unseen from a catalog by mail. The catalogs of the 1870s were commonly hardcover books put together by suppliers and manufacturers. These survive today in ever decreasing numbers and the pricing on these continually goes up.

A word of caution is necessary here concerning paper products. Most of these items of the post-Civil War era were made using acid-based paper. This means that they are slowly going to self-destruct because of the effects of the acid process. The best thing we can recommend is to have a valuable paper item such as a map professionally preserved using acid free backing and non-glare glass covers. Never but never expose old paper items to direct sunlight, as this will accelerate the destruction of the paper.

The most desirable catalogs are those from heavy equipment manufacturers who would have produced a minimal number of catalogs to sell their products. Some manufacturers did not produce catalogs that we know of, such as the famous Virginia and Truckee Railroad. This famous short line railroad also produced many items in its shops and foundry in Carson City, Nevada, but did not catalog anything that we know of. Basically they custom made whatever the client wanted.

Checks

In the days of the mining boom there were a surprising number of checks written in the course of business in the mines. In fact, banking of the time period to just before World War II was largely limited to those individuals and companies with a lot of money. The average individual did not put money into banks, but mines that were incorporated or large enough had to have a banking relationship somewhere. The problems this caused were enormous. Take for instance a company mine or mill in Treasure City, Nevada. Treasure City was located nearly three hundred miles from the state capital of Carson City, which was effectively the nearest rail terminus where mail could be delivered reliably to a bank in Carson or to San Francisco. Imagine the problems associated with sending a shipment of gold to the mint in Carson or San Francisco with the proceeds to be paid by check to an account in the Bank of California in San Francisco or Virginia City. What is a miner to do with a piece of paper in a mining camp when nobody will cash his check so he can buy provisions to make it through the winter?

The obvious answer is that there were not a lot of checks used during the time period up to say about World War I. That a number of canceled checks do exist today is evidence that such items were kept for long periods of time. Many of the period checks found today will be drawn on banks that no longer exist and this merely adds to the curiosity of such items. Values can be enhanced considerably if the checks are made to or by some notable, such as Samuel Clemens (a.k.a. Mark Twain) or his brother, Orion Clemens. Condition as always is highly important.

Deeds

Deeds are the sort of paper collectible that seem to have a way of hanging around in out-of-the-way places, such as trunks and closets. Consequently, they do not seem to draw the interest of many collectors unless they are particularly attractive or involve a notable personage. Occasionally they can have real import, such as when the presence of a man in a town is open to speculation and the evidence of a signed deed is uncovered.

Government Publications

From the earliest of times our national government was interested in what went on in various parts of our country. To document the expansion of the United States, the government produced highly detailed publications illustrating the many mining fields in the west, as well as some in the east. The western mines were the focus of some truly large maps and folios in the post-Civil War west when the country was expanding at an awesome rate. Books of one size or other were written and published covering exploration of most parts of the west, including Alaska. The early explorers were commonly in the employ of the federal government, either as regular US Army or as civilian employees, and normally wrote of their exploits that were then printed by the US government printing office. Individual agencies also printed books and later maps. All of these are more or less rare and the value will depend primarily on condition, with the exception of maps. Remember that bureaucracy always prints more than they think is necessary and as they do not have to make a profit on the book, frequently the age of a government printed book is not a good indicator of its value.

Script

This is an area of collecting that sometimes falls in the cracks and gets lost. Script as we know it here was paper money issued by mining companies or by local stores in mining camps to take the place of actual currency. Because of the remoteness of many western mining camps small fractional currency in the form of coinage was sometimes simply just not available. Paper script was issued to allow a storeowner to make change for small purchases when he did not want to extend credit. It was also a handy way to assure the return of the customer, since script was not legal outside the establishment where it was issued.

Collectors of coins and currency will claim script as their item for valuation and collecting, but older mining town script is highly valued by mining collectors.

Stock Certificates

Stock certificates issued from mining companies are some of the most sought after pieces of paper associated with the mining era. Collectors of stocks should probably only concern themselves with older issues and we will therefore only illustrate pieces issued before World War I. Initially stock certificates were rather small, being little larger than a business check of today. Indeed, it is possible to overlook a valuable stock certificate when digging through a trunk of paper by thinking of the larger 8" x10" fancy certificate of today.

The earliest certificates and those with the most value will come from mines that were in the western states. Not that the east did not have such items, but the wild speculation of the western mining boom was fueled by tons of paper stock certificates. Many of these older pieces of paper still occasionally surface when an old trunk is rooted out from its resting place in some dusty attic. The illustrations in this book will cover a cross section of certificates from the oldest and rarest to the common but still collectible.

Stock certificates from eastern coal mining regions have attracted little interest in recent auctions, probably because of the quantity available. Larger size certificates with ornate engravings will bring substantially more in price than plain certificates. Visual quality is extremely important. Certificates that have been canceled through a sale or retirement have less collector interest than issued and uncanceled sheets. Folding or mutilation, or any other defacing of a certificate all detract from the value of the certificate. In some other cases the signature of some dignitary is what will add value to an otherwise common certificate.

Mining Magazines

An often overlooked area of collecting in the mining field is old mining magazines. These trade papers have real interest, but are not universally valued by collectors, with some rarer issues bringing as much as ten times the price of other contemporary issues.

An early lithograph of Virginia City, Nevada, mines. $1,000 to $1,500. RB

Advertising pamphlet from J. W. Killinger of Lewiston, Idaho. $100 to $125. RB

Mining journals in good condition have become quite collectible. Pricing depends heavily on condition and geographic orientation. Little interest is shown for eastern mining papers, while early western journals have lately gotten expensive. $5 to $40 per copy. RB

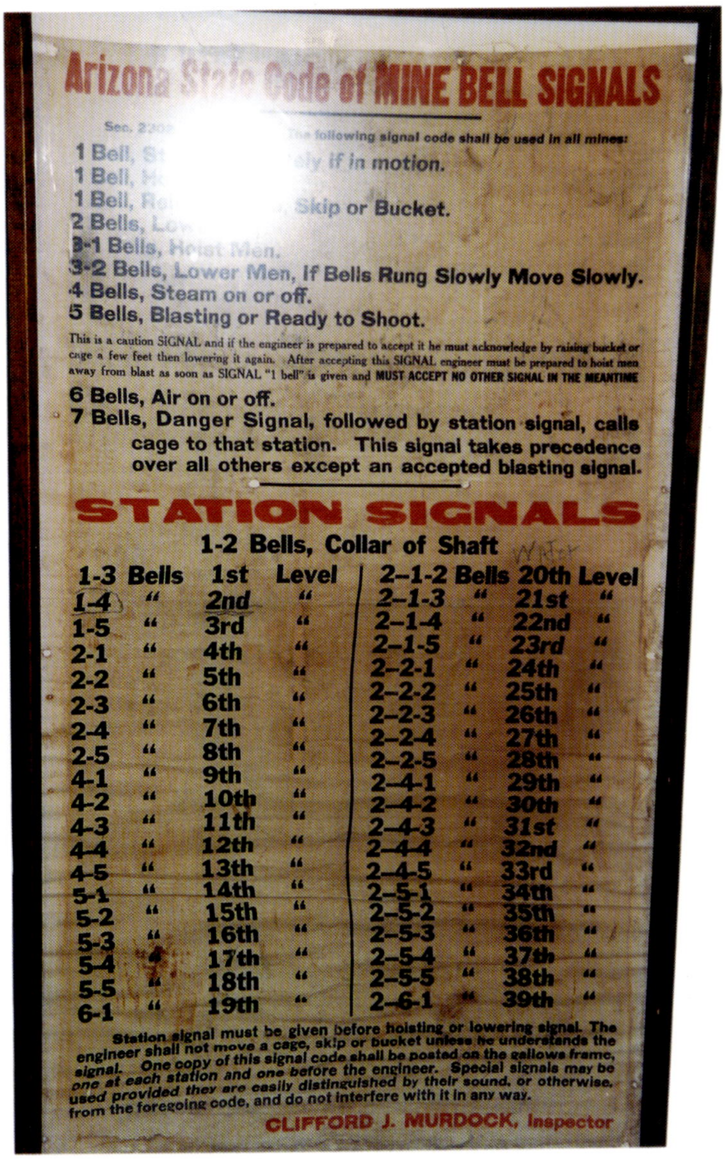

Documentation from mines or assayers is difficult to value. Condition is paramount to value, as is the obscurity of the mine or district. $100 to $150. RB

Left:
A framed and mounted poster detailing the sequence of mine bells in use in Arizona. Much like other states, Arizona standardized the bell system for informing miners where a mine elevator was going and what was happening elsewhere in the mine. $250 to $400. SB

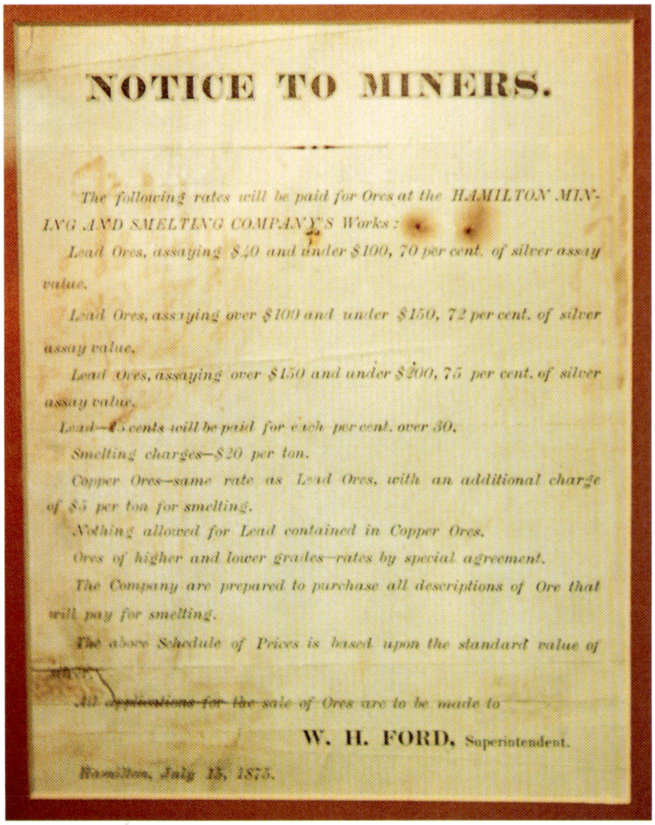

NOTICE TO MINERS.

The following rates will be paid for Ores at the HAMILTON MINING AND SMELTING COMPANY'S Works:

Lead Ores, assaying $40 and under $100, 70 per cent. of silver assay value.

Lead Ores, assaying over $100 and under $150, 72 per cent. of silver assay value.

Lead Ores, assaying over $150 and under $200, 75 per cent. of silver assay value.

Lead—[illegible] cents will be paid for each per cent. over 30.

Smelting charges—$20 per ton.

Copper Ores—same rate as Lead Ores, with an additional charge of $5 per ton for smelting.

Nothing allowed for Lead contained in Copper Ores.

Ores of higher and lower grades—rates by special agreement.

The Company are prepared to purchase all descriptions of Ore that will pay for smelting.

The above Schedule of Prices is based upon the standard value of

All applications for the sale of Ores are to be made to

W. H. FORD, Superintendent.

Hamilton, July 15, 1873.

A printed "Notice to Miners," this is an offer to buy ores in the Hamilton, Nevada, area by the Hamilton Mining and Smelting Company works. Date of the notice is July 15, 1873. The Hamilton area was notable for the extreme richness of its silver ore and the rates quoted on this notice bear this out. Hamilton is now a ghost town and accessible by twelve miles of bad dirt road. $300 to $550. RB

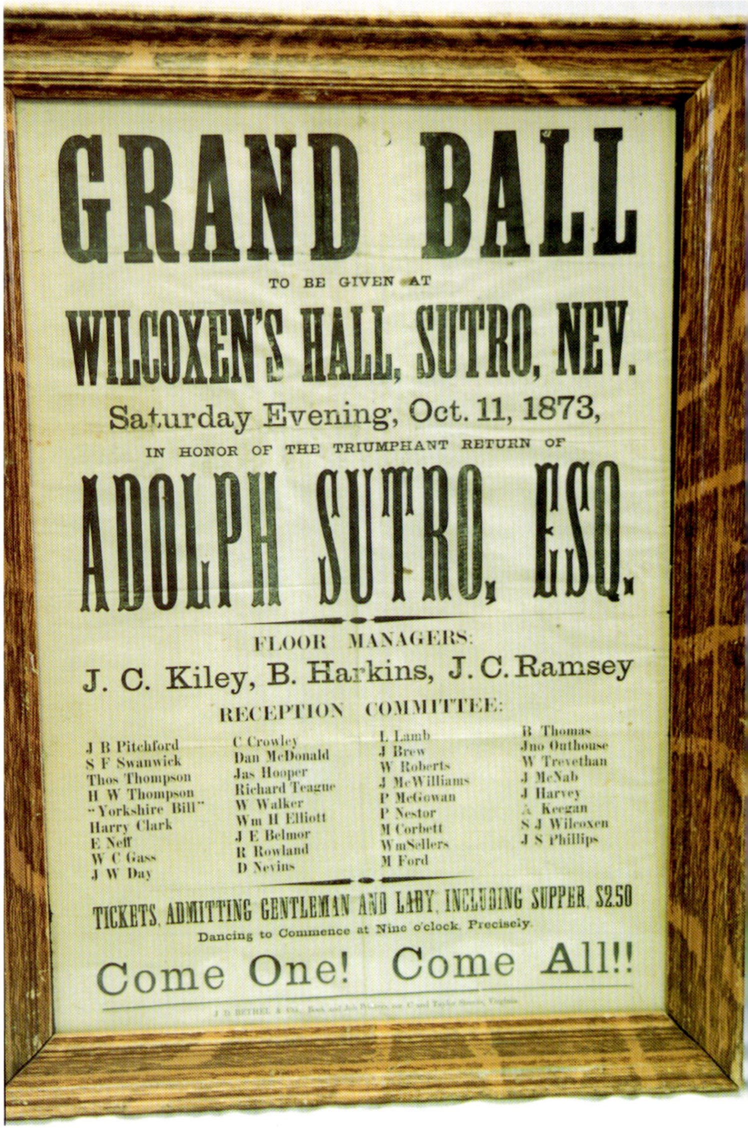

GRAND BALL

TO BE GIVEN AT

WILCOXEN'S HALL, SUTRO, NEV.

Saturday Evening, Oct. 11, 1873,

IN HONOR OF THE TRIUMPHANT RETURN OF

ADOLPH SUTRO, ESQ.

FLOOR MANAGERS:

J. C. Kiley, B. Harkins, J. C. Ramsey

RECEPTION COMMITTEE:

J B Pitchford	C Crowley	L Lamb	B Thomas
S F Swanwick	Dan McDonald	J Brew	Jno Outhouse
Thos Thompson	Jas Hooper	W Roberts	W Trevethan
H W Thompson	Richard Teague	J McWilliams	J McNab
"Yorkshire Bill"	W Walker	P McGowan	J Harvey
Harry Clark	Wm H Elliott	P Nestor	A Keegan
E Neff	J E Belmor	M Corbett	S J Wilcoxen
W C Gass	R Rowland	WmSellers	J S Phillips
J W Day	D Nevins	M Ford	

TICKETS, ADMITTING GENTLEMAN AND LADY, INCLUDING SUPPER, $2.50

Dancing to Commence at Nine o'clock Precisely.

Come One! Come All!!

Right:
A framed poster advertising a "Grand Ball" on October 11, 1873, in Sutro, Nevada. Now just a memory, the Sutro tunnel still exists and was intended to drain the water from the mines of the Comstock Lode in Virginia City. Sutro was a master promoter and this poster would have considerable value today. $1,500 to $2,000 RB

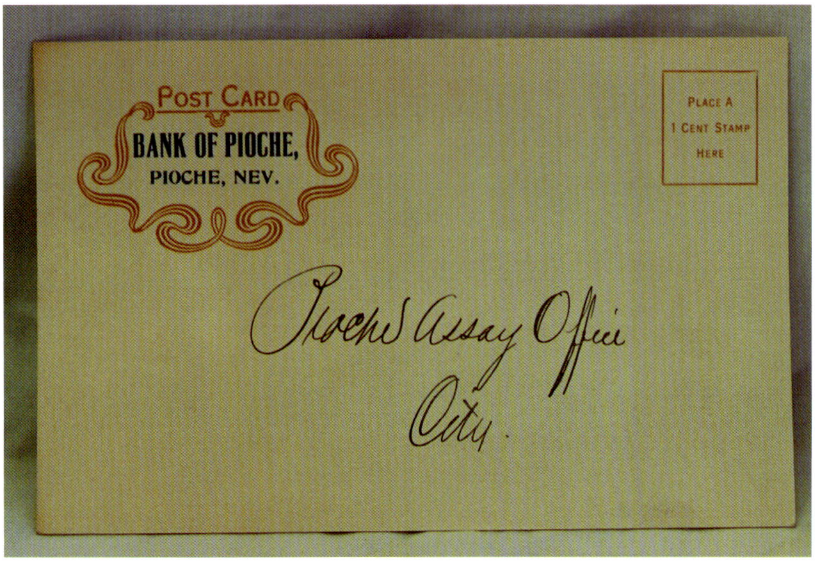

A post card addressed to the Pioche Assay Office in Pioche, Nevada. Pioche still exists approximately 100 miles to the north of Las Vegas with many old buildings and mining mementos to be observed and purchased. $40 to $65. RB

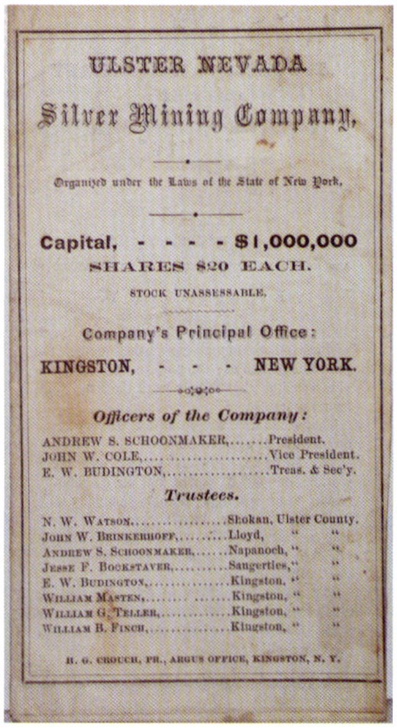

ULSTER NEVADA

Silver Mining Company,

Organized under the Laws of the State of New York,

Capital, - - - - $1,000,000

SHARES $20 EACH.

STOCK UNASSESSABLE.

Company's Principal Office:

KINGSTON, - - - NEW YORK.

Officers of the Company:

ANDREW S. SCHOONMAKER,.......President.
JOHN W. COLE,.......Vice President.
E. W. BUDINGTON,.......Treas. & Sec'y.

Trustees.

N. W. WATSON,.......Shokan, Ulster County.
JOHN W. BRINKERHOFF,.......Lloyd, " "
ANDREW S. SCHOONMAKER,.......Napanoch, " "
JESSE F. BOCKSTAVER,.......Saugerties, " "
E. W. BUDINGTON,.......Kingston, " "
WILLIAM MASTEN,.......Kingston, " "
WILLIAM G. TELLER,.......Kingston, " "
WILLIAM B. FINCH,.......Kingston, " "

H. G. CROUCH, PR., ARGUS OFFICE, KINGSTON, N. Y.

Prospectus for the Ulster, Nevada, Silver Mining Company from a group out of New York. No reference to Ulster, Nevada, could be found by this author. $150 to $300. RB

A paper sign from an assay office in Rochester, Nevada. $200 to $300. RB

Stock certificate from the Silver Peak Valcalda Gold Mining Company dated 1908. This area is a remote and very rugged mountain range which is still occasionally mined. $50 to $100. RB

A very simple certificate from the Treasure Hill area of east central Nevada. The area was a classic boom and bust district when the high-grade silver ore petered out a few yards below the surface of the ground. It didn't help that the mountainous area is snowbound and very cold during the winter. $200 to $400. RB

Unused envelope from the S.E. Simmons Mfg. Co. The Simmons company made carbide miners' lamps. $25 RB

A large mounted canvas poster from Bureau of Mines, State of Colorado, detailing the various bell signals in use in Colorado mines. Not dated but probably pre-1910. $300 SB

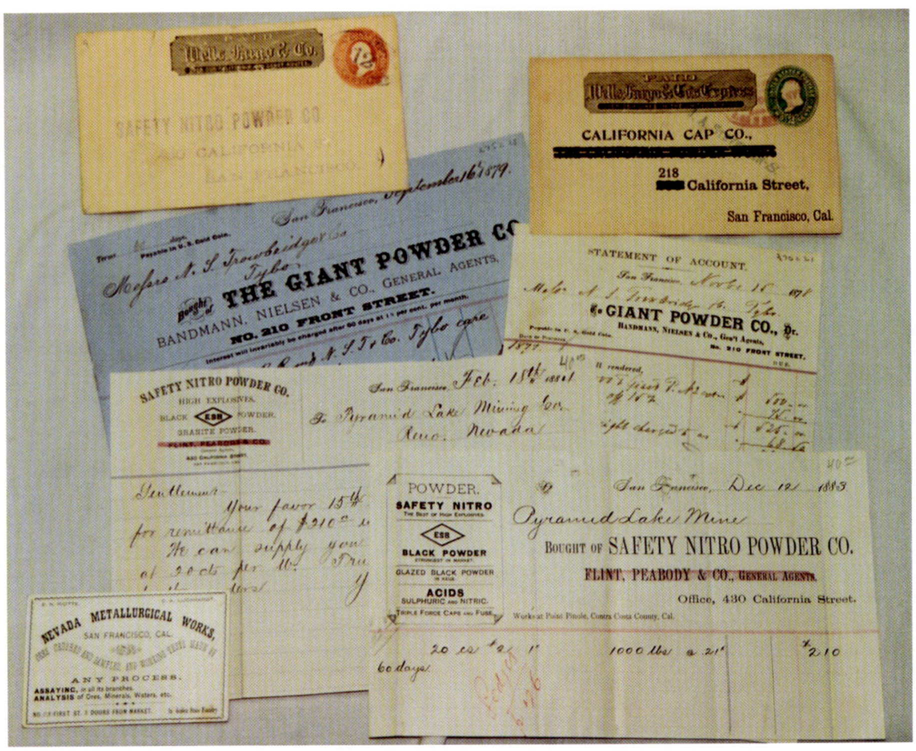

Six paper items to or from western powder companies. $30 to $100. RB

Four small sample lists of prices for mining equipment. $150 to $200. RB

Common paper items from or to mine suppliers do not generate high prices unless they contain the signature of some notable individual or refer to a well known company. $5 to $100. RB

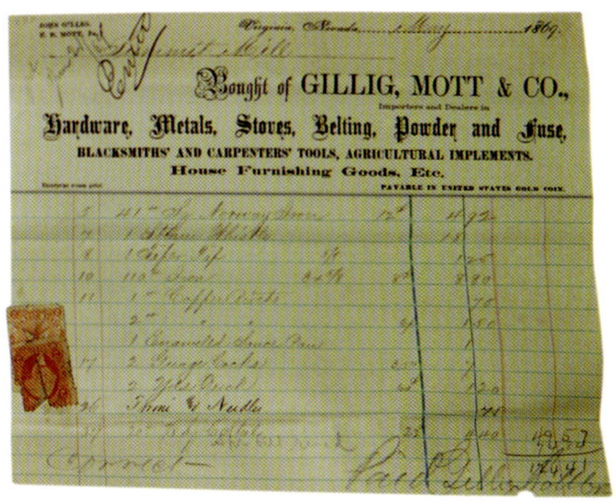

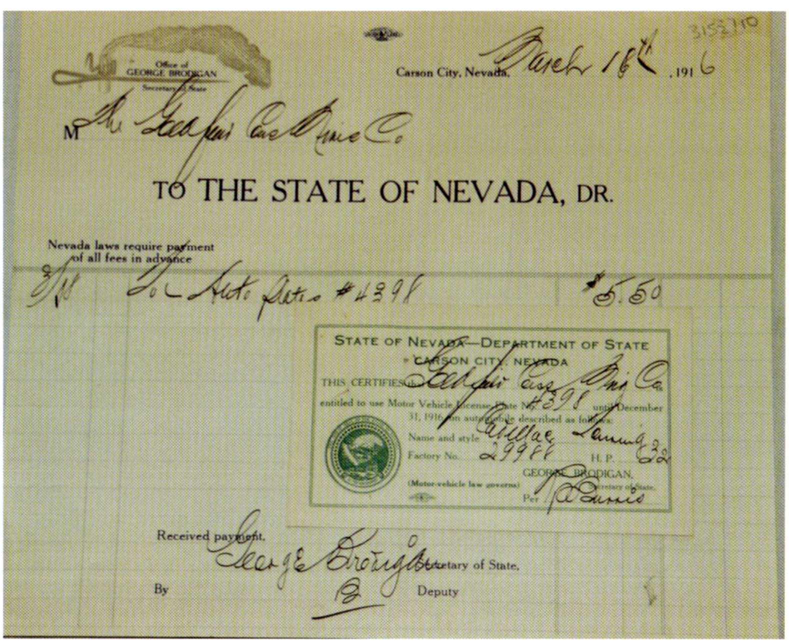

Check from the Bank of Rawhide dated 1908. Rawhide was one of the wildest and most famous gold rush camps in Nevada. $25 to $50. RB

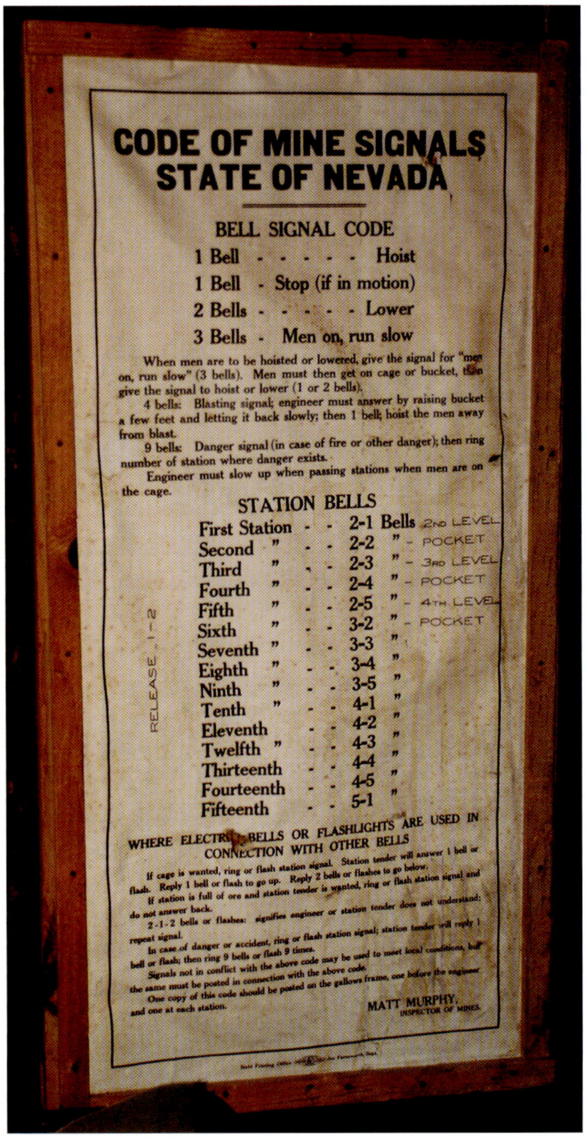

A paper poster of the mine bell signals from Nevada is a good find, as most signs would have deteriorated in wet conditions. This one is mounted on a board to preserve it. $500 to $800. RP

Now badly faded, this original US patent was issued to a Mr. John Crause of Virginia City, Nevada. $100 to $200. RB

Revenue bond issued by the Virginia & Truckee Railroad Company of Carson city, Nevada, and never recalled. The V & T was considered the "Queen of the short lines" and anything marked with V & T has a premium value. $300 to $500. RB

Stock Certificate from the Homestake Mine of South Dakota. Cancellation date is July 1883. Values depend on who signed the certificates. $25 to $150. RB

Very early stock certificate from a location west of Reno and near the California line. Dated 1864. $350 to $600. RB

Another very fancy stock certificate from West Virginia, this time dated 1901. Coal mining stocks do not bring high prices. $5 to $20. RB

Dated 1868, the Continental Consolidated Gold and Silver Mining Co. was located in the mountains near the hamlet Markleeville, California, which still exists. $350 to $800. RB

Stock certificate from the Tamarack and Custer Consolidated Mining Company of Carson City, Nevada. Issued June 1913. Note the fancy engraving. $10 to $25. RB

Stock certificate dating from January 1864 from the Pennsylvania mine of Michigan. $25 to $40. RB

Check from the Miners and Merchants Bank of Mina, Nevada, dated 1903. The town of Mina, Nevada, still exists as a slow spot on highway 95. The bank is long gone. $25. RB

Stock certificate from a mine in the Territory of Montana dated 1886. $60 to $80. RB

Very fancy stock certificate from Tonopah, Nevada, dated 1916. Note the embossed gold seal at left and the engraved scenes, all of which add value. $20 to $40. RB

Historic stock certificate from the "Battling Nelson Mining Company," named after the loser in one of the big fights held in Nevada during the first part of the twentieth century. $60. RB

Stock certificates from Mexican mines are very rare and this one is dated August 22, 1865, making it a very early example from "Lower California." $300. RB

The Homestake Mining Co. still exists in one form or other and still produces gold in the Dakotas. This certificate dates from 1913. $10 to $25. RB

Actually a piece of sheet music, this is chiefly of interest because of the photos of the lost mining town of Rawhide, Nevada. The area is now a strip mine hole in the ground. $350. RB

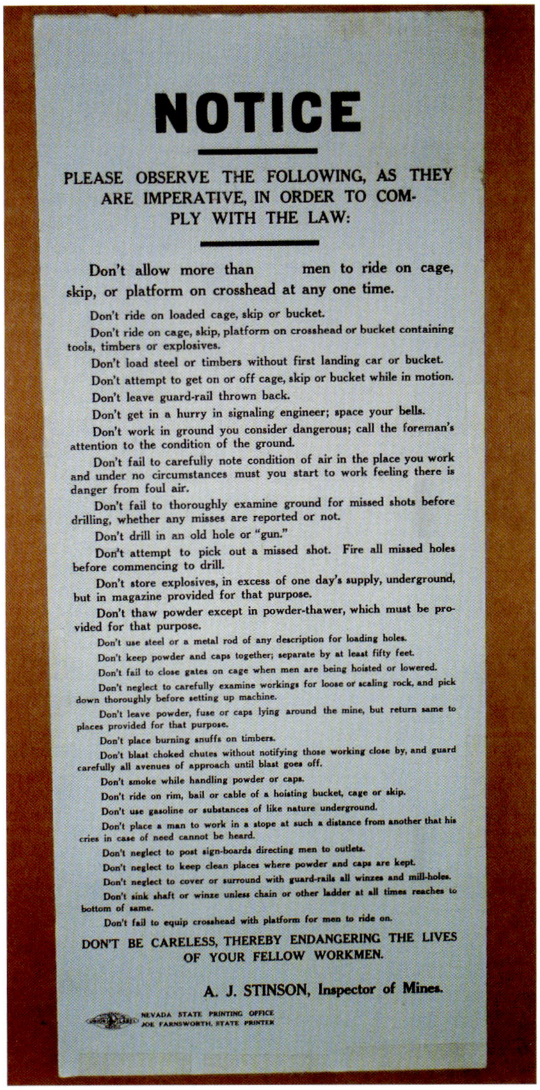

NOTICE

PLEASE OBSERVE THE FOLLOWING, AS THEY ARE IMPERATIVE, IN ORDER TO COMPLY WITH THE LAW:

Don't allow more than men to ride on cage, skip, or platform on crosshead at any one time.

Don't ride on loaded cage, skip or bucket.

Don't ride on cage, skip, platform on crosshead or bucket containing tools, timbers or explosives.

Don't load steel or timbers without first landing car or bucket.

Don't attempt to get on or off cage, skip or bucket while in motion.

Don't leave guard-rail thrown back.

Don't get in a hurry in signaling engineer; space your bells.

Don't work in ground you consider dangerous; call the foreman's attention to the condition of the ground.

Don't fail to carefully note condition of air in the place you work and under no circumstances must you start to work feeling there is danger from foul air.

Don't fail to thoroughly examine ground for missed shots before drilling, whether any misses are reported or not.

Don't drill in an old hole or "gun."

Don't attempt to pick out a missed shot. Fire all missed holes before commencing to drill.

Don't store explosives, in excess of one day's supply, underground, but in magazine provided for that purpose.

Don't thaw powder except in powder-thawer, which must be provided for that purpose.

Don't use steel or a metal rod of any description for loading holes.

Don't keep powder and caps together; separate by at least fifty feet.

Don't fail to close gates on cage when men are being hoisted or lowered.

Don't neglect to carefully examine workings for loose or scaling rock, and pick down thoroughly before setting up machine.

Don't leave powder, fuse or caps lying around the mine, but return same to places provided for that purpose.

Don't place burning snuffs on timbers.

Don't blast choked chutes without notifying those working close by, and guard carefully all avenues of approach until blast goes off.

Don't smoke while handling powder or caps.

Don't ride on rim, bail or cable of a hoisting bucket, cage or skip.

Don't use gasoline or substances of like nature underground.

Don't place a man to work in a stope at such a distance from another that his cries in case of need cannot be heard.

Don't neglect to post sign-boards directing men to outlets.

Don't neglect to keep clean places where powder and caps are kept.

Don't neglect to cover or surround with guard-rails all winzes and mill-holes.

Don't sink shaft or winze unless chain or other ladder at all times reaches to bottom of same.

Don't fail to equip crosshead with platform for men to ride on.

DON'T BE CARELESS, THEREBY ENDANGERING THE LIVES OF YOUR FELLOW WORKMEN.

A. J. STINSON, Inspector of Mines.

NEVADA STATE PRINTING OFFICE
JOE FARNSWORTH, STATE PRINTER

Large paper sheet notice of the mine rules from an early mine in Nevada. $150 to $200. RB

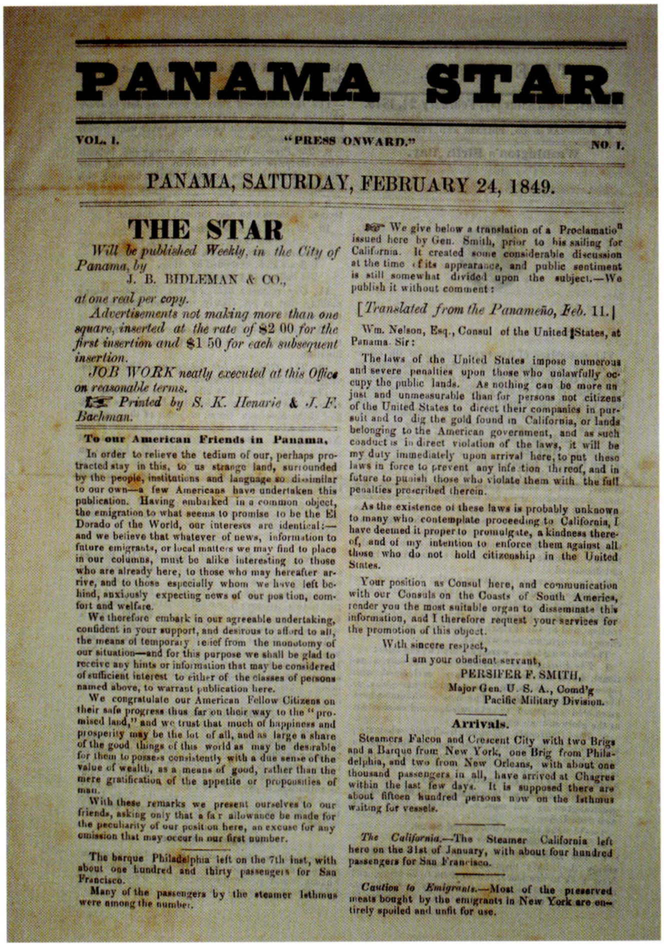

PANAMA STAR.

VOL. I. "PRESS ONWARD." NO. I.

PANAMA, SATURDAY, FEBRUARY 24, 1849.

THE STAR

Will be published Weekly, in the City of Panama, by

J. B. BIDLEMAN & CO.,

at one real per copy.

Advertisements not making more than one square, inserted at the rate of $2 00 *for the first insertion and* $1 50 *for each subsequent insertion.*

JOB WORK neatly executed at this Office on reasonable terms.

☞ *Printed by S. K. Henarie & J. F. Bachman.*

To our American Friends in Panama,

In order to relieve the tedium of our, perhaps protracted stay in this, to us strange land, surrounded by the people, institutions and language so dissimilar to our own—a few Americans have undertaken this publication. Having embarked in a common object, the emigration to what seems to promise to be the El Dorado of the World, our interests are identical:—and we believe that whatever of news, information to future emigrants, or local matters we may find to place in our columns, must be alike interesting to those who are already here, to those who may hereafter arrive, and to those especially whom we have left behind, anxiously expecting news of our position, comfort and welfare.

We therefore embark in our agreeable undertaking, confident in your support, and desirous to afford to all, the means of temporary relief from the monotomy of our situation—and for this purpose we shall be glad to receive any hints or information that may be considered of sufficient interest to either of the classes of persons named above, to warrant publication here.

We congratulate our American Fellow Citizens on their safe progress thus far on their way to the "promised land," and we trust that much of happiness and prosperity may be the lot of all, and as large a share of the good things of this world as may be desirable for them to possess consistently with a due sense of the value of wealth, as a means of good, rather than the mere gratification of the appetite or proponsities of man.

With these remarks we present ourselves to our friends, asking only that a fair allowance be made for the peculiarity of our position here, an excuse for any omission that may occur in our first number.

The barque Philadelphia left on the 7th inst, with about one hundred and thirty passengers for San Francisco.

Many of the passengers by the steamer Isthmus were among the number.

☞ We give below a translation of a Proclamation issued here by Gen. Smith, prior to his sailing for California. It created some considerable discussion at the time of its appearance, and public sentiment is still somewhat divided upon the subject.—We publish it without comment:

[*Translated from the Panameño, Feb.* 11.]

Wm. Nelson, Esq., Consul of the United States, at Panama. Sir:

The laws of the United States impose numerous and severe penalties upon those who unlawfully occupy the public lands. As nothing can be more unjust and unmeasurable than for persons not citizens of the United States to direct their companies in pursuit and to dig the gold found in California, or lands belonging to the American government, and as such conduct is in direct violation of the laws, it will be my duty immediately upon arrival here, to put these laws in force to prevent any infection thereof, and in future to punish those who violate them with the full penalties prescribed therein.

As the existence of these laws is probably unknown to many who contemplate proceeding to California, I have deemed it proper to promulgate, a kindness thereof, and of my intention to enforce them against all those who do not hold citizenship in the United States.

Your position as Consul here, and communication with our Consuls on the Coasts of South America, render you the most suitable organ to disseminate this information, and I therefore request your services for the promotion of this object.

With sincere respect,

I am your obedient servant,

PERSIFER F. SMITH,

Major Gen. U. S. A., Comd'g

Pacific Military Division.

Arrivals.

Steamers Falcon and Crescent City with two Brigs and a Barque from New York, one Brig from Philadelphia, and two from New Orleans, with about one thousand passengers in all, have arrived at Chagres within the last few days. It is supposed there are about fifteen hundred persons now on the Isthmus waiting for vessels.

The California.—The Steamer California left here on the 31st of January, with about four hundred passengers for San Francisco.

Caution to Emigrants.—Most of the preserved meats bought by the emigrants in New York are entirely spoiled and unfit for use.

A complete copy of the *Panama Star* newspaper written and printed in Panama for early gold seekers going to the fields by boat to Panama, then across the isthmus, then by boat up the coast to California. Quite rare. $3,000 to $6,000. RB

Right:
Some later mines used a phone bell system for communication instead of bells. This paper mounted sign is such an item and is rare because they normally did not last well. $250. HD

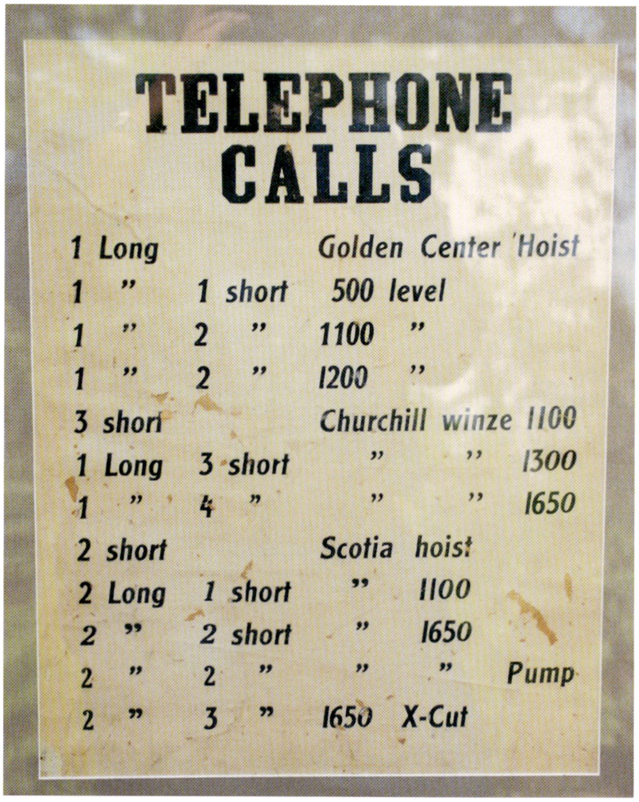

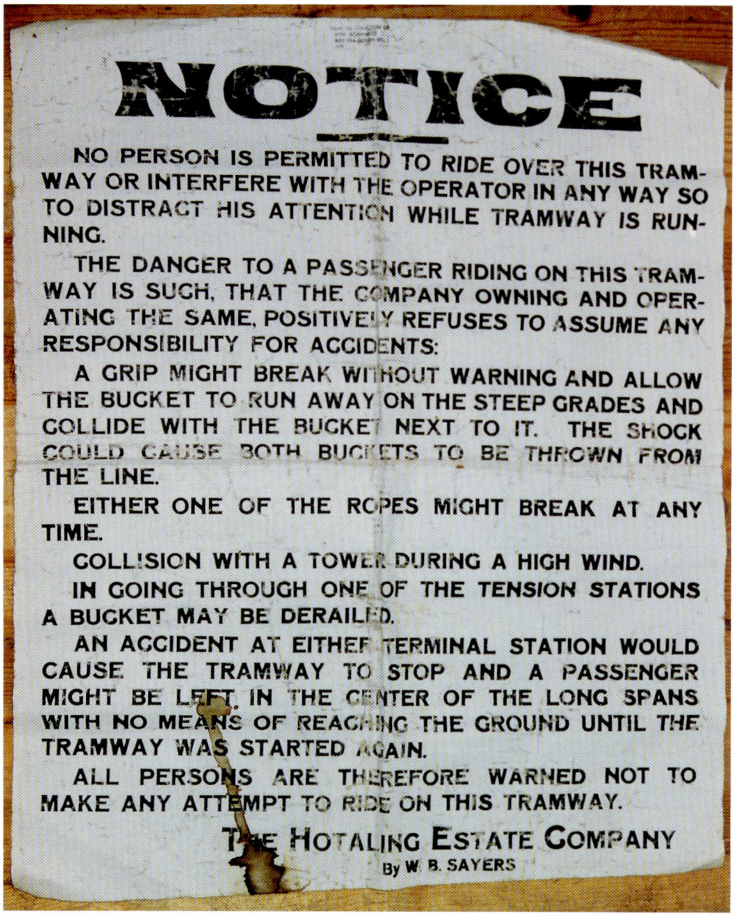

Poster prohibiting men from riding bucket tramways. This dangerous practice was not encouraged. Tramways were used to move quantities of ore from mines to a mill or shipping point in many areas. One can still be seen going over the highway in Pioche, Nevada. The poster is mounted on linen and is about twenty-four inches by thirty inches. $350 to $500. RB

This large poster is in mint condition and comes from Butte, Montana. Date is unsure. $600. HD

A Labor Day picnic sign from Grass Valley, California. Dated at post-1894 because of the holiday designation. About $600. HD

A piece of script of company money issued by the Manhattan Silver Mining Co. of Austen, Nevada. These were never issued, as the government got wind of their printing and put a stop to the practice before it began. RB

Another company store booklet of script from the copper mining town of Ludwig, Nevada. This was another attempt to alleviate the shortage of hard currency in the mining towns of the west. RB

Chapter Twenty-one

PHONES AND COMMUNICATION EQUIPMENT

We do not often think of communication equipment in mines as something worth having, but the fact is miners from the very earliest times needed a better way to communicate with each other than the human voice. The earliest form of communication and one of the most effective was the use of bells. At first the bells were actuated by mechanical means, usually using long ropes. Later on electricity came into use to ring bells and this proved so successful that a standardized system of rings was developed to signal miners what was going on. This system was very much the same from one state and mine to another, as indicated by the photos of bell signs in this book. The actual bells are not remarkable and can be found in many places in the country.

As soon as the telephone came into being mines adopted their use for talking between the surface and the lower reaches of a mine. While phones were important to miners, the use of bells and horns continues to this day in mines.

An intact Edison wet cell battery. Used inside mines to provide electrical power, these batteries were dangerous to the user. $75 to $100. RB

An early electric storage battery case. These would have acid in the glass, with the lead plates suspended inside and the connections on top. These were used inside mines to power lights and phones, as well as bells. $50 to $75. RB

Four different mine bells: three are manually operated, with the one in top center electrically activated. $50 to $125. RB

Phones from the first were adapted to use in mines. Wooden-framed models did not hold up to the abuse of miners and companies like Western Electric developed phones like this cast-iron-cased model. $350. HD

Chapter Twenty-two

PHOTOGRAPHS

The collecting of old photos is so large an area of mining collectibles that we can scarcely touch the surface in the space available in this book. Still, we need to make the collector of mining gear aware that early photos of many mining towns, camps, and the mines themselves are in some cases extremely valuable. In many cases the photograph will be the only evidence of what may have been a thriving community and will be of interest to historians as well as collectors.

For those people who spend their off hours actually finding old towns, sometimes a photo with a distinguishing landmark will be the difference in locating the right spot. We know, for example, of more than a few towns that no longer exist where the location can only be positively identified by early photos. Historical preservationists frequently buy photos of old buildings to restore them to their former glory.

In this same area of collecting, old glass plate negatives can be invaluable because they are commonly the only trace of early photos where the actual prints have long ago disappeared. Condition of prints and/or negatives is all important.

An old photo of the "Miners Bar." Location is unsure, but probably somewhere in Nevada. $150. RB

In early western mining communities there was usually a town photographer who could provide individual portrait work or sell you a "house" photo, one which he took and sold prints of. Called by some "cabinet" or stock photos, this one is of interest because it pictures the group of miners who escaped a fire in one mine by tunneling through to another nearby mine shaft. It is from Virginia City, Nevada. $650. RB

Another stock photo of the Hale & Norcross mine of Virginia City, Nevada. $650. RB

An individual portrait using a staged set to allow the intrepid miner to send a memento back east to his family. $600 to 800. RB

Another smaller type of common portrait photo, this one is from Virginia City, Nevada. Photos from certain locales typically bring more on the market than others. Virginia City is one of the highest value photo producers. $400. RB

A field photo of a small mining operation, probably from Tonopah, Nevada, post-1905. This photo is mounted on a heavy cardboard backing and may have been sold that way. $300. RBc

Another cabinet or stock photo, this one is of more interest because of the subject matter. It shows the items produced by the Fulton Foundry of Virginia, Nevada (Virginia City), sometime in the 1870s or 1880s. $650. RB

Of interest because of the location, this photo is of Silver Plume, Colorado, and was done by a Denver photographer. $275. RB

Another cabinet photo of two green-horn prospectors and their trusty burro headed out for fame and fortune. The photo was taken by Luke and Wheeler and is assumed to be from Colorado. $300. RB

Another cabinet photo, this time of a firewood supplier in the Sante Fe area. We assume this to be New Mexico but are not sure. $150. RB

There are still some of us old enough to remember stereopticons and the fantastic photos. The stereo photos were published by many photographers and publishers and were hugely popular. This pair is of the Coeur D'Alene area in Idaho and would appear to be very early in that area's development. $200. RB

This photo gives details of the mining cars built at the Fulton Foundry, including the weight of the car and its load capacity. $650. RB

Chapter Twenty-three

PROSPECTING GEAR

Prospecting gear can be very basic and for purposes of this book we will include only two types of items: DRY WASHERS and GOLD PANS. Gold pans are rather obvious, but early pans are not quite so apparent. Do not be fooled by fakes in this area. The single piece cold-formed steel pan is a late development and no matter how rusty and authentic that pan looks, unless it was put together out of several pieces of steel it probably is not real old. Look for hand soldering and edges that are not rolled.

Dry washers have been around since the earliest times and were used whenever a prospector found a likely stream bed in a desert environment. Lacking water to operate a sluice box or gold pan, the dry washer came into use. This mechanical device was operated by moving air and agitating sand which may or may not contain fine gold. The process moved the sand through a system of screens using air and motion and hopefully collected the gold present. Some worked, some didn't, but they are an interesting idea and intact dry washers are highly collectible in today's market.

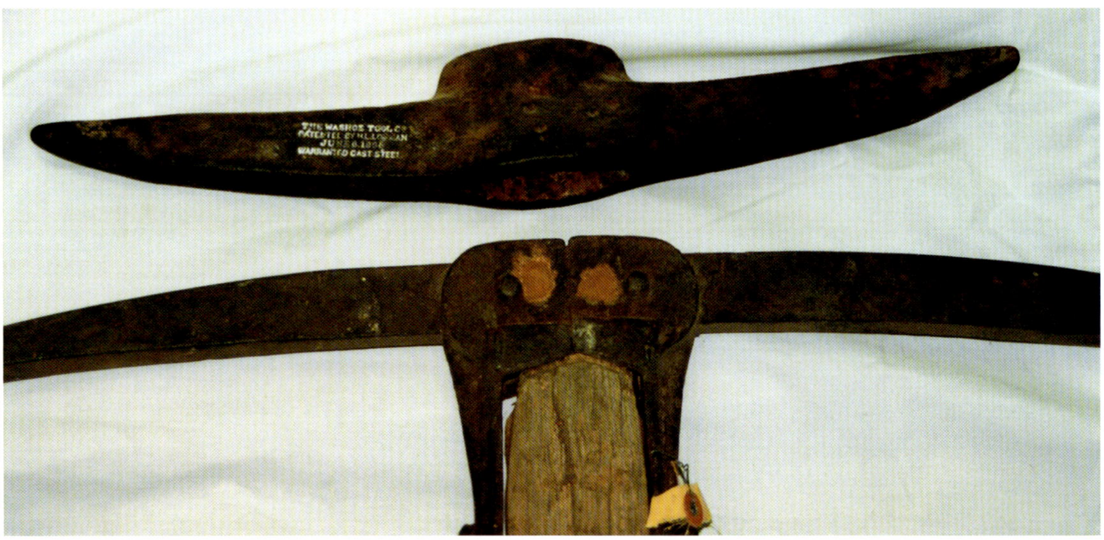

Miners' picks were pretty plain items, but every once in a while someone tried to improve on the standard form. Both of these were marked with patents, while the bottom one is designed to be folded for carrying. These date from the 1860s. Top $350, bottom $200. RB

Not all mining collectibles are neat packaged items. This obviously hand-forged tool was probably used as a starter drill when two miners were beginning a hole in rock for blasting. It could have also been used as a pick. Value is pegged at $25 by its owner. Items such as this can easily be faked so beware. RB

Very early gold pans were made from sheet tin or steel and were soldered together. This one dates from the mid-1800s and is in good condition, about fourteen inches across. $140 to $180. RB

Below:
These four instruments are all "pocket" items for miners. The top right is a barometer that is missing its leather top cap. The other three are compasses and could have been used during prospecting or in laying claims where some precision was required. The large compass at top left, $70 to $175. RB

Real gold pokes are difficult to establish without documentation. These pokes are all of deer hide and probably date from the late nineteenth century. The object facing into the one poke is a horn spoon and was used like a funnel to put fine gold into the pouch or poke. $50 to $300, depending on condition. RB

A selection of old miners' picks. Plain, $20 to $50. Engraved or fancy ornate pieces, $350 to $400. EC

Known as a "dry washer," this device was used to process small quantities of gold-bearing sand or dust. The material to be processed was put into the top while the crank was turned. The mechanical action moved air through screens to the right while agitating the whole affair. It worked, but only as well as the user could make it work. Rare in any condition. $900 to $1,400. NM

Above and left:
A very good-condition "dry washer." The detail photo shows the pulley works on the inside. Not very big, these items are very rare and seldom found with the feeder bins and the crank system intact. The crank on this one is made of small threaded pipe and is probably not original. $1,000 to $1,500. RB

Gold pokes, all old. Depending on age, condition, and authenticity, $50 to $150. RB

A hand-held lope-engraved "Los Angeles, California," for examining ore. $35. RB

A small low-power microscope for examining ore samples. $100 to $150. RB

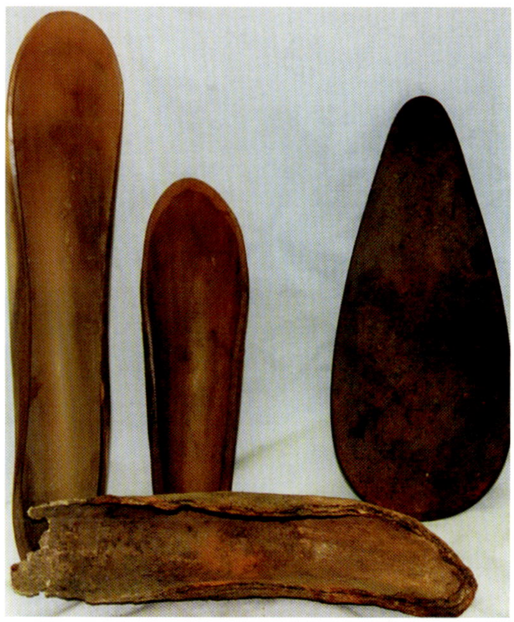

Horn "spoons" for handling gold dust. $90 to $125 each. RB

Pickaxes, $100 to $165.

Chapter Twenty-four

SAFETY LAMPS

Safety lamps are just what their name says: lamps that are safe to use in a possibly flammable environment. Safety lamps can be found with candles, oil, or carbide illumination. All are more or less collectible. The theory and practice of the safety lamp was that an open flame behind a metal screen would cause any errant gas to burn off in small amounts against the screen. In practice this would cause the screen to throb with small bursts of flame, as the methane present in a mine would burn off. When this happened the savvy miner would quickly leave the area, as it was almost a sure bet that sooner or later the gas buildup would result in an explosion.

The usage of safety lamps was much more common in the eastern coal mines than in the western metal mines, but miners are miners and many easterners who came west brought their tools with them, including their safety lamps. The majority of safety lamps were manufactured in England, but they have been made throughout the world. Many sizes exist and they can more or less be classified by the number of hours the lamps would burn on a single charging of oil or carbide. Photos of a variety of sizes are included in this section, including one with four different sizes for comparison. We have observed samples as small as three inches in height and as big as fourteen inches long. Workmanship in general is quite high on safety lamps and survival rate is rather good.

Safety lamps are more common in eastern mines than in the west and more common in Europe than in the US. In this photo the farthest left lamp is actually a carbide safety lamp while the others are all oil or alcohol fired. From left: $800, $350, $500, $1,500, $1,500, and $250. EC

Left:
Safety lamps were more common where explosive gases were present and many were produced overseas. From left: Baby Wolf, $250; unknown maker, $200; unknown maker, $200; and Patterson, $275. EC

Chapter Twenty-Five

SCALES

Assaying and Laboratory Scales

This type of scale is familiar to anyone who has had a high school chemistry course or been in contact with an assayer. These are precision instruments—even old ones—and they should be treated with respect. Older scales would normally be contained in a square wooden and glass-sided case and include a set of weights and pans for weighing very small quantities of metals. Even when they have been used for years these scales are accurate and can be quite valuable. They have been made in a variety of sizes through the years and this writer had the privilege of operating a scale for several years capable of handling one thousand ounces. That scale was also accurate enough to weigh a cat's whisker!

Pocket Scales

Pocket scales were meant to be carried in a jacket pocket inside a case of metal—usually tin—with a small set of weights. These scales were either held by hand using a string or attached to a stand that was placed in a hole in the case. The early ones were basically all the same, varying only in size and the quality of adornment on the case. They are highly collectible these days and condition is paramount, as is the completeness of the set. Weights should be present, as they are frequently lost in part or in whole. The buyer should be very aware that fakes are still being manufactured in China today and these are cheap and virtually worthless. We have observed them for sale for as little as $14.95.

Weight Sets

In the process of doing assays, various weights were used to counterbalance the ore sample being tested. These weights were precise and in troy ounces and grains and so marked. Most commonly they would be made of brass or copper to avoid corrosion and adulteration. Larger weight sets up to one thousand ounces were also used and some of these would be made of iron. Complete sets of weights are not too common, as the smaller pieces tend to become lost in usage. Larger single-iron weights will occasionally be found, but they are of interest only as a curiosity. Complete sets will bring rather higher prices than those missing individual pieces. Early weight sets were frequently kept in fancy containers and at least one example of an ivory carrier exists.

Adulterated sets of weights are not unknown and are highly sought after by collectors. The adulterated sets will typically have a small hole in the base of the brass weight, which would have been filled with a heavier metal such as lead. The owner of the weights would then weigh out a portion of gold that was actually heavier than it should be and the buyer got more gold than he should have.

"Miners Improved Gold Scale" pocket scale is original and in near fine condition. The case is painted tin with the words "Manufactured expressly for California" printed at the bottom. $400 plus. RB

Pocket scales made for the Alaska trade are scarce. These pocket scales are two different sizes. $1,500 each. EC

Pocket miner's scale "made expressly for California." Items such as this were intended to catch the eye of the early forty-niners. The "made for" label was also used for the Alaska gold rush many decades later. $600 and up. RB

A pocket miner's scale for rough weighing of gold. This one is in a leather case. Marked "Miners Companion," $2,000+. RB

Two weight sets: the one on the left has no weights but is made of ivory and marked. The one on the right is complete but plain. About $100 each. RB

Portable scale set complete with weights. $200 to $250. RB

This assayer's scale is unusual, since it was made from cigar boxes and is entirely hand-built. $300 to $400. RB

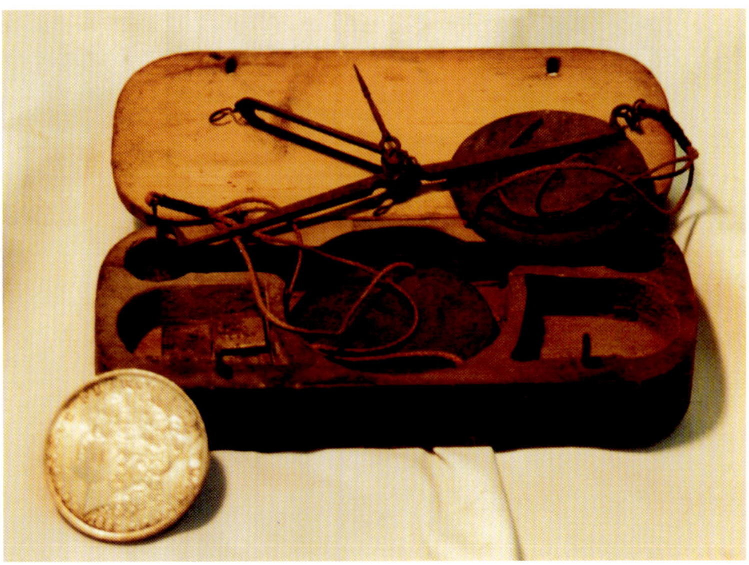

This rather crudely made scale was designed to be held by one hand in operation. It is quite small in size, as can be seen by the silver dollar alongside. The carrying case is also hand-made of wood with no embellishments whatsoever. Set with the wooden box with weights, $100 to $200. RB

Above and left:
Miner's pocket scale probably intended for the Alaska trade, as it is painted for the Seattle Hardware company. Complete and in good condition. $750 to $1,000. BM

A small pocket scale from San Francisco manufactured for the "low end of the market." It came in a cardboard box with instructions for its use printed in the top. It does not appear very sturdy or well made and was apparently not a commercial success, as it is quite rare now. $200 to $350. AM

Chapter Twenty-six

SPOONS

Spoons for the purposes of this book can be divided into two categories: commemorative and usable. The commemorative spoons are for all intents no different than those collected today. They were sold and given as mementos throughout the west and east and some were remarkably beautiful, as can be seen in the photos accompanying this section. Most will be found in silver, with some plated in gold or with gold accents. Enameling and engraving all add value and interest to the collector. Spoons were commonly given to girls when they turned a particular age—usually sixteen, although we have seen spoons with names and ages engraved with much younger ages.

The usable spoons with which we are concerned here are of a much different type and were made out of a variety of materials, including polished horn. These spoons were used for handling of gold dust to avoid the possibility of catching a few flakes under a fingernail of a bartender or assayer weighing out an ounce to pay his bar tab. Whatever the spoon is made of it had to be able to freely give up gold dust with a shake.

The commemorative spoons have been selling for some rather high prices as of this writing and quality is the most important factor.

Four souvenir spoons of silver with gold plate. All are from Colorado. $50 to $150. RB

Gold-plated collector's spoon $75. RB

Silver collector's spoons. About $125 each. RB

A fine-condition spoon engraved "Tonopah, Nevada." Gold on silver. $130 to $160. RB

A collector spoon made of silver with a miniature windlass and gilt plating on the spoon. $300 to $400. RP

This silver spoon is complete with a windlass and hollow handle that actually allows the crank to be turned and the bucket raised and lowered to the miner engraved on the spoon. Possibly unique. $300 to $400. RP

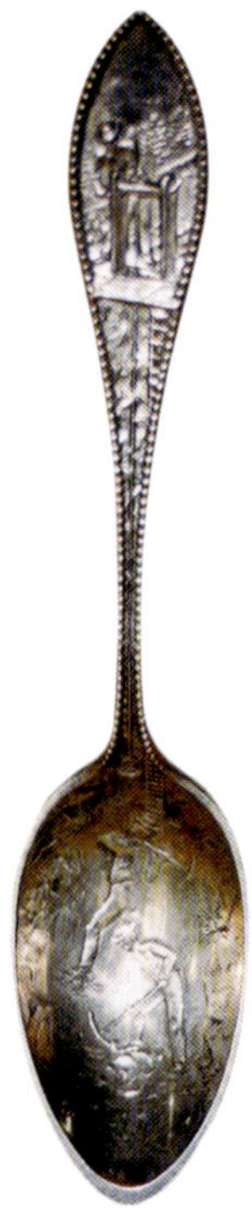

Another finely crafted silver spoon with miners engraved on all sides. $175. RP

Chapter Twenty-seven

STRONGBOXES

Strongboxes are the stuff that legends are made of, as well as Hollywood scripts. In the west and much of the east, a good strongbox was the next best thing to a heavy safe to keep a precious metal shipment in the hands of its proper owners. Today, the real strongbox is a highly collectible item, especially if it has Wells Fargo & Co. painted on its side. The fact that everybody used strongboxes does not account for the relative scarcity of the items today. Virtually every mine had one in addition to a safe, as did freight offices, stage lines, mail carriers, and anybody else who had a valuable stash to hide.

Antique safe from the 1880s or 1890s. $800 in average condition. RP

Every mine or transportation office had a strong box to hold gold or specie until it could be sent to the mint or given out to employees. This one happens to have been used at some time by the Railway Express Agency. RB

Made of heavy wood with metal edges and a padlock for security, this is typical of a mine's strongbox security. RB

Chapter Twenty-eight

SURVEYING EQUIPMENT

Surveying equipment is a frequently overlooked piece of collectible mining equipment. Few people think of surveying as part of mining, but if it were not for highly adept surveyors many mines would have fallen into disuse very quickly. As an example of how precise the old time surveyors were we only have to point to the famous Sutro Tunnel, which was punched through a mountain to help drain rivers of water from the Comstock Lode of Virginia City, Nevada. When one realizes that the tunnel was dug uphill through the base of a mountain five miles away from the shafts of the Comstock and broke out into the precise shaft it was planned to intersect one can understand the abilities of the old time surveyor.

In addition to their ability, these old timers used pencil-driven math and super high quality transits to achieve their ends. When mapping out mine shafts these craftsmen did not just measure using a tape, they used highly specialized equipment adapted to underground mining. Transits that could sight three angles at once were used, as well as precision plumb bobs with oil lamps in them (to be seen in the darkness of a shaft). These are all now antiques for the collector as global positioning satellite (GPS) systems have taken over. The finest of this surveying equipment will be in very good condition and normally be housed in some sort of hard case for carrying and storage. Hard cases can be either wood (preferred) or leather. Where optics are present they should be clear and in good condition.

Since some of these items are not quite museum quality (but close) pricing is difficult to pin down. Essentially the market will dictate what these tools are worth.

Finely made "dip needle" with box, used to "find ore bodies." In fine condition and quite rare. $150 to $175. RB

An impressive surveyor's transit, this fine condition tool is also out of the Nevada State Museum's collection. It is capable of measuring in three modes at once. $5,000. NM

Set of two plumb bobs with wicks for use inside mines when surveying. Quite rare with the box in good condition. $1,900 to $2,500. RB

A finely crafted folding compass used in surveying. Lower serial numbers bring much higher prices. $100 to $450. RB

An anemometer used to measure air flow into or out of a mine. This instrument is complete with a carrying case of leather and is finely made. $150 to $175. RB

A mining clinometer for measuring the angle of a slope inside the mine. $300. AM

A rare item known as a Surveyor's Target by Buff and Buff. Not your everyday household item. $1,000. BU

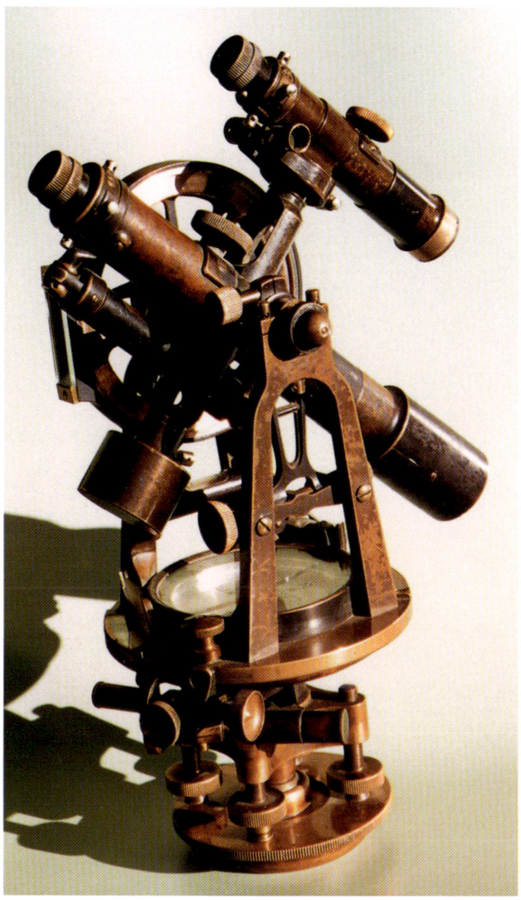

A mining transit by C.L. Berger & sons dating from 1915. $2,500 or more.

Chapter Twenty-nine

TOKENS

Other types of tokens were used during daily operations and these are samples of common ones. Early underground miners were made to pay for explosives and supplies used using tokens. At the end of the week the miner was paid after the cost of his supplies were deducted. This helped cut down on the theft of explosives and perhaps made the miners a bit more efficient in their use. $2 to $20. RB

Tokens almost fall into the area of numismatic collecting, but there have been some tokens and tool tags that were almost unique to the mining industry. Most all of the tokens or chits were used to note the presence or absence of a particular tool from a locker. In practice, the worker who was checking out the tool to be used in the mine would take the brass tag and place it on a board covered with pegs. When he returned the tool the token was placed back in the "in" position.

The other use of chits or tokens was for the miner to move a specifically numbered token from the inside of a numbered pegboard to the outside, indicating that the miner had left the insides of the mine. This was a safety practice to prevent accidents when "shooting" or using explosives. Mines could be noisy places—even underground—and typically miners did not have watches to tell them when their shift was up. Blasting or shooting was commonly done when the miners were on a lunch break or when the shifts were changing. This practice evolved to reduce the number of men in the vicinity of the explosion and cut down on accidents. The tokens are normally rather crude and do not display minting as would a coin. Frequently they would have been made up by the mine's tool shop, which would stamp them out using sheet brass and letter dies.

Condition when found will almost always be used, but in otherwise good condition. Values are not high in most cases.

Tool tokens or "chits" are less common than they used to be. Not rare, they are commonly faked, so the collector should be wary of such items unless documented. $10 to $30. DS

Chapter Thirty

EPHEMERA

There are a number of other collectibles that more or less defy categorizing and this chapter is the catch-all for those goodies. Sometimes the items described and shown here will have considerable value. This is not meant to denigrate the importance of these items, but for this book we can't find a good niche to put them in.

A very early molded bottle from Eureka, Nevada. $250 to $400. RB

Milk or cream pitcher from the Independence Mine of Cripple Creek, Colorado. $100. RB

Another photo of the inside of an assay shack showing the delicate assay scale to the right and a less precise pulp scale to the left rear.

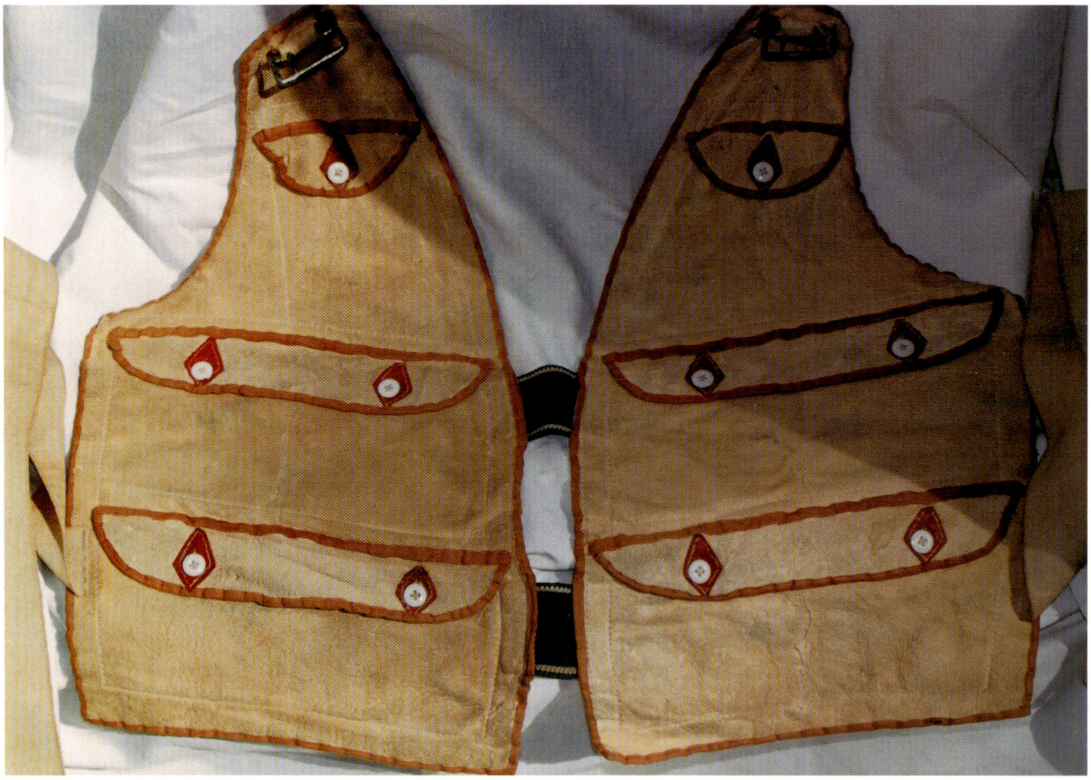

Very rare, this item is a money vest made of deer skin and fabric. In use, the returning gold rich forty-niner would carry his gold coins in the pockets and wear the whole affair underneath a shirt so it could not be seen. Early miners regarded paper money as totally unreliable and took gold coin whenever possible. $1,800 to $2,200. BM

Of little collector value, these are original stick matches with strike anywhere sulfur tips. They came in bricks packaged in boxes and can be very flammable. Value is nominal. RB

An original prescription bottle from Tonopah , Nevada. $200 to $250. RB

A pile of lunch pails gathering dust. EC

These miniatures are not toys, but are in fact examples of the square set timbering process developed in the Comstock mines and later used world wide. Thought to be engineer's working examples, these are quite rare and very fragile. $500 to $750. RB

This coin was minted in January 1867 from the "First product of the mill process in the Pah-Ranacat Mining District, Nevada." $300 to $400. DS

A complete brick of "Comet Matches." $100 to $125. RB

A small portion of one man's collection of mining books. EC

The "Klondyke Nugget Game" is very rare and dates from the turn of the nineteenth century. $1,000. EC

Commemorative plates are a later development, but this one is from early Butte, Montana. $150 to $250. RB

We were very fortunate to snap this shot of a large banner from the Gould and Curry mine of Virginia City, Nevada. This flag is approximately six feet by eight feet and languishes in a basement storage room of the Nevada State Museum. $6,000 to $10,000. NM

Corporate seals are like walls: always there but unnoticed until you need them. These are all old and from defunct mines of Virginia City, Nevada. $70 to $350, depending on condition and ornateness. RB

rather nice inkwell that is silver
o date. $125. RP

eft:
andle holder from
e in the town of that
100. RB

Stained glass window from the house of a long-gone mine owner. The crest is finely done and in an arch pane. It was not for sale! RP

Trophy or desk ornament. Whatever its original purpose, it is cast of heavy metal with traces of gilding. $150. HD

This large silver-plated trophy comes from Butte, Montana, and is dated 1926. Given out by a medical supplies maker, it is finely made. $200 or more. HD

This is a small water flask from the What Cheer Tool Co. of Iowa in new condition. $225. HD

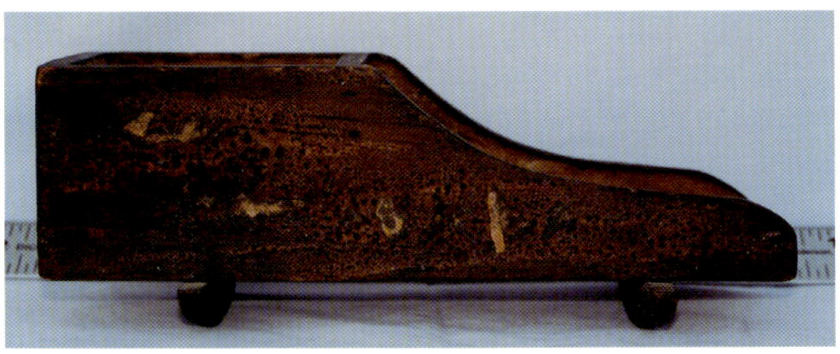

A miniature rocker (gold panning device); this cute little item is old, but its value is minimal. RB

These two bronzes stand about twelve inches high and are of lightweight casting. Probably made in France, they are finely crafted. $250 each. HD

Miner's shaving mug. $250-$350. RS

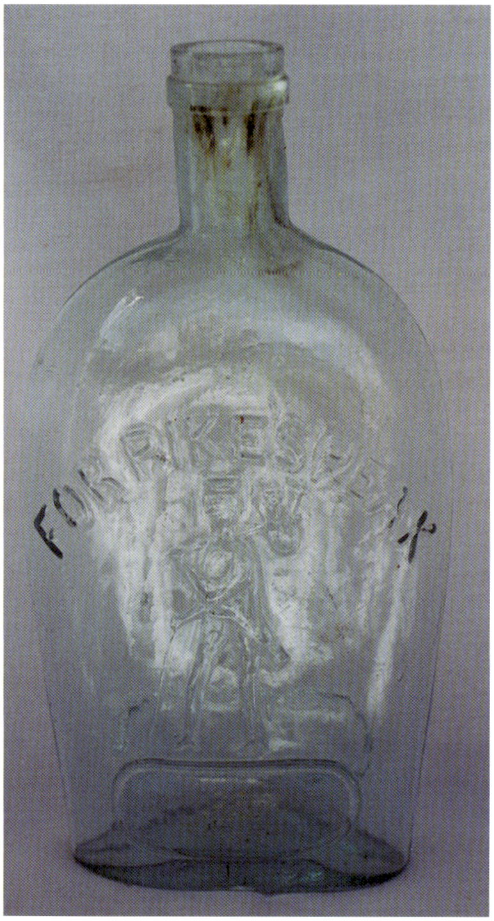

A very early bottle with the words "for Pike's Peak" cast into the glass with the figure of a miner. No definite date, but probably 1870s or earlier.

Local brewers created some interesting labels for today's collectors to find. This one came from Goldfield, Nevada, and is dated after 1906. It is one of only two known. RB

Two hand-made mining souvenirs from Butte, Montana. The shovel is dated 1932. $250 plus. AM

BIBLIOGRAPHY & REFERENCES

Books

Bailey, Lynn R. *Supplying the Mining World, The mining Equipment Manufacturers of San Francisco: 1850-1900*. Tucson, Arizona: Westernlore Press, 1996.

Clemmer, Gregg S. *American Miners' Carbide Lamps*. Tucson, Arizona: Westernlore Press, 1997.

Paher, Stanley W. *Nevada Ghost Towns & Mining Camps*. Las Vegas and Reno, Nevada: Nevada Publications, 1970.

Pearson, David W. *This Was Mining in the West*. Atglen, Pennsylvania: Schiffer Publishing, 1996.

Ramsdell, Jack and Wagner, Norman. *Patents Miner's Candlesticks*. Carson City, Nevada: self published, 1982, 1993.

Sagstetter, Beth and Bill. *The Mining Camps Speak*. Denver, Colorado: Benchmark Publishing of Colorado, 1998.

Temple, John. *Mining, An International History*. New York: Praeger Publishers, 1972.

Twitty, Eric. *Blown to Bits in the Mine: A History of Mining & Explosives in the United States*. Ouray, Colorado: Western Reflections Publishing Company, 2001.

Wilson, Wendell E. and Bobtink, Ted. *A Collector's Guide to Antique Miner's Candlesticks*. Tucson, Arizona: Mineralogical Record Inc., 1984.

Periodicals

Collector's Mining Review, The, various issues. Reno, Nevada: Lane Griffin.

EUREKA! The Journal of Mining Collectibles, various issues. Peoria, Arizona: Dave Thorpe.

Mining Artifact Collector, various issues. Tucson, Arizona: Wendell E. Wilson.

Other References and Sources

Ron Bommarito
P.O. Box 1,
Genoa, Nevada 89411
775-782-3893

Errol Christman
P.O. Box 313
Cedar Ridge, California 95924

Fred Holabird
Holabird Mining, Environmental & Historical Consultants
3555 Airway Dr #308
Reno, Nevada 89511
775-852-8822
fred@holabird.org

David W. Pearson
LSB Books
P.O. Box 381
Genoa, Nevada 89411

Roger & Connie Peterson
The Miner's Pick Antiques
14207 Highway 49
Amador City, California 95601
minerspick@goldrush.com
209-267-0848

INDEX